GOVERNMENT GIRL

GOVERNMENT GIRL

YOUNG AND FEMALE
in the
WHITE HOUSE

Stacy Parker Aab

ecco

An Imprint of HarperCollins Publishers

ecco

An Imprint of HarperCollins Publishers

REMOVED FROM COLLECTION

I have changed the names and identifying features of some individuals to preserve their anonymity. Portions of dialogue were taken from my journal entries of the time period; other conversations were remembered to the best of my ability.

HarperCollins books may be purchased for educational, business, or sales promotional use. For information, please write: Special Markets Department, HarperCollins Publishers, 10 East 53rd Street, New York, NY 10022.

FIRST EDITION

Designed by Suet Yee Chong

Library of Congress Cataloging-in-Publication Data has been applied for.

ISBN: 978-0-06-167222-4

10 11 12 13 14   OV/RRD   10 9 8 7 6 5 4 3 2 1

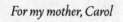

For my mother, Carol

CONTENTS

ॐ III ॐ
AFTER

Down for the Night

Okinawa, Japan
2000

THE DOOR TO THE PRESIDENT'S SUITE STOOD STRAIGHT AHEAD and marked the end of my path. Few staffers had business being inside the president's suite. I did. I was the presidential advance person in charge of the hotel.

The president had traveled to Okinawa, Japan, for the 2000 G8 summit. This was the last evening of the trip, and the president and staff were still out for dinner. Down the hallway I went, walking past the rooms of key staffers, each door posted with a paper sign bearing his or her name. Doug Band, the body aide. Sandy Berger, the national security advisor. Donnie Flynn, the deputy special agent in charge (DSAIC). We labeled doors so staffers could find one another easily, per presidential trip protocol.

The president's door remained unmarked. An agent stood posted outside, the only visual anyone needed to know which room was his. The agent smiled as I approached.

I wanted to check the suite pantry. The hotel treated us superbly, so I had little doubt that all of the sodas and snacks had been refreshed. We stayed in a resort hotel, the ANA Manza, that rose up like a white ship on a slip of beach peninsula. Step outside, and you were surrounded by the healing beauty of blue waters, by the air blanched with light and heavy with moisture. Before this trip, I had known Okinawa as the site of the bloodiest battle in the World War II Pacific theater, the Japanese and American forces sustaining more casualties than the civilians in Hiroshima and Nagasaki combined. I had not known that they had slaughtered each other in paradise.

Before arrival, I worked with the hotel and the Secret Service to prepare the president's suite. I arranged for and assigned staff rooms, making sure that every person who deplaned Air Force One, and each support plane, had the right place to sleep—no easy feat, given shifting manifests and sold-out hotels. Other advance people had finite assignments. They finished their events, and they would be "down"—done for the day and not expected to be at anyone's immediate beck and call. Not me. As the hotel advance person, known as the "RON," for "remain overnight"—shorthand for the advance person in charge for wherever the president spent the night—I was always on, from the minute the president and his staff arrived till the minute they were wheels up on Air Force One. RONs worked long hours, but I was proud to be part of a small group of advance people considered qualified for the task.

The posted agent let me into the suite. Lamps were glowing, as were overhead lights. I could not see how dark it was outside, for each pair of curtains was closed. But the rooms felt bright, regardless, with their white walls and caramel wood accents. Windows lined the bayside and oceanside walls, and when the curtains were open, the water dazzled and the air swept through the dining room and sitting room in a perfect cross-breeze.

"You must hate that there's balconies," I had said to my RON agent counterpart on our first walk-through.

"If I have my way," he said, "he'll never know there are balconies."

I wondered if he was joking or if he planned to sew the curtains shut. We both knew that the president had reacted badly when the RON agent had done that on the South American trip a few years prior. "It looks like a goddamned cave in here!" he had reportedly howled. In minutes, the agents had unstitched the curtains. No one wanted to be yelled at by the president, but that was the risk we took when we negotiated the line between security and suffocation. Still, the president took his own risks when he resisted the agents' measures: they kept curtains shut to keep possible snipers from getting a line of sight.

The agent and I walked out onto the balcony. Below lay the lagoon, where toddlers were wading with their young parents. From the dining room balcony the president could look out onto the sea itself. A Japanese navel vessel sat out there, fat and unmoving, securing the site. Waiting, watching.

"I'm closing these curtains," my counterpart said. With one strong motion, he shut out the sky, the water, and all of that light.

"You're not going to pin them shut now?" I asked.

"No," he said.

"Thank you. You know he doesn't like that."

But later, when I inspected the suite before the president's first arrival, I saw safety pins glinting in the sheers and I swore out loud. My agent counterpart had pinned the sheers shut. *Pinned.* Maybe he hadn't done hotel work for a while or his supervisor had told him to do it and he had had no choice, but we all knew that the president did not like to be so obviously closed in. The audacity, really, to pin shut the curtains of the president of the United States.

We would let the supervisors fight it out. I knew from

experience that I would not be the one unpinning the curtains if the president roared. It would be the agent.

And it was. The pins came out as soon as the president arrived.

Now, two days later, I checked the refrigerator to see if anything needed restocking. The valet managed this now, but I kept tabs too. I was happy to find it full of Diet Coke, water, and plates of plastic-wrapped vegetables and crudités. Happy until I realized that I was hungry. Did I have time to run to my room before the president returned from his dinner? Yes.

The fruit basket Mr. Abe sent to my room every morning was there, filled with the most luscious fruit I had ever tasted. Mr. Abe was the Tokyo hotel manager entrusted with running the site during our stay. I was his advance staff link to the president, and he treated me well: the mangoes oozed juice at a mere touch. If I were still White House staff, I'd have to worry about violating ethics rules by accepting a gift that might exceed the $20 gift allowance, given how much the hotel might charge for such a basket. But I was ex-staff. A volunteer, from the trusted pool of people the White House asked to do occasional advance work. I could eat the mangoes.

I rushed downstairs and ate the fruit on my balcony, careful to keep the juice off my suit. The president was due back soon. I kept the radio bud planted in my ear, listening for his departure from the dinner site. I could see the Japanese vessel, large and imposing on the waterline. Later, I would arrange for Mr. Abe's senior executives to be photographed with the president. He would tell me that this honor was so appreciated that one of the executives invited him for golf—a major gesture in their corporate culture. Mr. Abe's happiness over this reward would become one of my favorite advance trip memories. People loved this president so much, and we did what we could to facilitate these moments of connection.

Eagle, depart. Eagle, depart.

Arrival pending! These were *my* moments of excitement, every time the president arrived or departed my site. I rushed to the lobby

to wait near the foot of the red carpet, the only staffer expected—
and allowed—to wait near the president's entrance point into the
hotel. The bellman would run the vacuum quickly over the carpet,
to make it fresh. I could step on it first if I wanted to, in those delicious moments we anticipated his arrival, but I never did. The first
shoe indents should be his.

People stopped. They watched this young woman of such a
mixed race that she could have been almost anything (Indian?
Native American? Louisiana Creole?), with dark hair just past her
shoulders, in a soft pink skirt, holding a clipboard and carrying a
black bag over her shoulder. *Look at that earpiece. She must be with
him.* At times, I knew people assumed I was Secret Service, and
if I told them to move, they would. But sometimes I wondered if
their imaginations wandered when they saw my youth and my fe-
maleness. *Does he hit on her, too?* We were only a year and a half past
impeachment and the failed Senate conviction. The president had
survived, even enjoyed sky-high approval ratings that last summer
of his term. But Monica was never far from our collective mind.

I tried not to worry about the assumptions of strangers. I had
learned quickly in Washington not to let others' weak moments
stop me in my tracks. The president had never behaved inappro-
priately with me, but plenty of other men in that ecosystem, from
journalists to staffers to Secret Service agents, had overstepped
the limits, forcing me to smile and laugh and pretend that nothing was wrong in order to keep on going. As a twenty-six-year-old
woman, I knew that men and women—including myself—made
mistakes. I could forgive that. I had been angry with the president
for what he had done with Monica, preimpeachment, but I was
also sick to death of what Mrs. Clinton called the "politics of personal destruction." The White House I had left in 1998 felt like a
frozen sort of hell, where we knew any words spoken or written
could be shared on demand with a grand jury. If anything, I was
a talker! I was used to trying to speak truth to those who did not

want to hear it. The White House in 1998 was no place for me.

But I still did occasional advance work, and I loved it. My anger toward the president had subsided, especially after the hypocrisy of his accusers was laid bare, including extramarital affairs by former Speaker of the House Newt Gingrich and the man slated to become speaker, Bob Livingston. I stood at the edge of the carpet feeling as if, despite having quit the West Wing the year before, I still belonged.

Agents and local security walked past as I waited in front of the second set of entry doors. In movies, you saw presidents enter buildings through back doors, kitchens, and hallways that moviegoers wouldn't know reeked of new paint and old room service. Our president didn't like this. He wanted his entrances to be as normal as possible. Luckily, many five-star hotels have VIP entrances that lead directly, discreetly to VIP elevators, allowing the feeling of normality (even if elite, monied normality) without the vulnerability of a main lobby route.

We tried to keep the president's entry area "clean," or as empty as possible. We never had the authority to shoo away guests, but often we did. The hotel agent did this. I found myself doing it, too, asking groups of people to move on. Sometimes overseas hotel staff and police officers forgot their roles and lingered, starstruck. If there were too many people around, instead of a quick, easy arrival, the president might get stuck shaking hands, giving autographs, and posing for photos. The president rarely seemed to mind, but this tacked on time and threw him off his already overpacked schedule. Staff like ourselves worked to keep him on task and protected from the reality that everywhere he went, crowds materialized.

"Five minutes out," my agent counterpart said.

I was present at arrivals because of protocol. I was there because the president of the United States and his staff *should* have a trusted staffer nearby who answered to him first, not to hotel management or agency directors.

I heard the pilot car siren wail. They were near.

We ran on adrenaline. Push forward. Push forward and engineer the best outcomes possible. After too many sixteen-hour days running up to Game Day, I needed to channel my inner long-haul flight attendant. Before each arrival I splashed my face and patted it down. I refreshed my eyeshadow and eyeliner and put on a coat of light lipstick: just because I was exhausted did not mean I had to look it. The sugar from the mango perked me up and I felt alert, even if my heels hurt from too much standing around in my black cowboy boots, boots I'd worn with certain skirts and suit slacks since my freshman year of college—the year I'd first begun working in the White House.

The pilot car drove under the portico, sirens off. Lobby onlookers crowded in, pushing into my peripheral vision. I turned and saw that some were even midlevel White House staff; no one seemed immune to the desire of watching or of just happening to be around and maybe attracting a president's hello, handshake, or conversation.

The motorcade cars curved in and came to a stop.

The hesitation, the all clear, before the special agent in charge (SAIC) stepped out of the limo's front passenger door. Usually this was the handsome, stoic Larry Cockell. I respected him so much, given all he had been through testifying before Ken Starr's Grand Jury regarding the president and former intern Monica Lewinsky, and how fatherly he had been to me on the road. But Cockell wasn't on this trip. In his stead was the DSAIC, Donnie Flynn, who had also persevered through those tough months, always with a smile and a kind word to share.

Doors had opened on all the other cars. Staff poured out, hustling, trained to hurry up and get out of the cars so as to enter a site when the president entered, and not be left out of his bubble. However, this was the end of the night. Only a few aides, such as the doctor, the military aide, and the body aide, needed to be with the president in the elevator ride to his floor. The others did not *need* to.

The DSAIC opened the president's door.

The president exited his limo when he was ready and not before. If he was speaking on the phone or with his wife, he kept talking. I did not know the signals that passed between the president and the agent in charge.

He exited the car. I waited until the president was near me before I moved. He made eye contact. I smiled, nodded, and turned around, beginning the walk to the elevators. Some staff had already walked past me. I "led" the president, even though, at this point, everyone knew the way. Sometimes the president walked close to me. Sometimes he talked with me. He was not talking with me this time. He was speaking to Doug Band.

Already the president's waiting elevators had the suction force of a black hole—anyone and everyone wanted in. I wished there were teams of research psychiatrists monitoring the movements and thoughts of the president's gaggle as they approached two open elevator doors. Most knew better than to get in with him if they did not need to. Some did not. Every elevator had its maximum weight-bearing capability. Too many people, and the elevator would creak and moan. Too many more, and the president's elevator car would stop between floors. This was our nightmare, so we tended to make up a manifest for each ride and print the manifests in the final schedules.

The problem lay with the egos and needs of those who orbited the president. You could be powerful in your own right, affecting national policy, shaping the legislative agenda, advocating for who was saved or who was annihilated by America's military might. This did not mean, however, that you needed to be at the president's side during an elevator ride, unless you were part of the "tight package" of security and critical staffers tasked to always be near him. People assumed that because they were high-ranking or just because they wanted to be near him, they could follow him into the car. Most of them were wrong.

As a RON staff advance, I had ridden in elevators with the president, but on this trip I was not manifested. There was not enough room. I did not *have* to be there. The RON agent rode with the president and the "tight package" and could lead them to the suite. I usually walked them to the held elevators, then race-walked to the normal guest elevators and caught up with them upstairs. This had been my routine for the last two days.

But sometimes the president messed up everyone's plans. That night, the president looked me dead in the eyes.

"You comin'?" he asked.

I did not answer. I just walked in, turned, and stood in front of him. I felt flattered as hell, but I tried not to show it.

No one dared question him. The elevator was packed, and my agent counterpart gave me a look. The elevator was manifested for eight, and now, somehow, there were ten of us. Didn't matter. POTUS had asked me in, and that was that.

The doors closed. Slowly. Then the elevator went shuttling straight up with the right speed and with no sense of strain. A male official laughed about something that had happened at dinner. I counted floors, praying the elevator would make it.

The car stopped at the president's floor and the doors opened. The gaggle headed down the POTUS hall, toward the suite and to their nearby rooms.

I would wait at the atrium railing until I received word that the president was down for the night.

Hearts? Did I hear someone say hearts? Would he while away his last night in Okinawa playing cards? Mrs. Clinton was not on this trip, but their daughter Chelsea was, keeping her own schedule. She was not back yet. I wondered if late-night card playing would keep her up. She did have her own room, adjoining but private. A card game was unlikely to disturb her, I figured. Yet the president entered the suite alone.

A passing agent told me that POTUS was down for the night. Good, I thought, though I'd wait until Doug gave me the all clear himself. I felt proud knowing that for forty-eight hours, the president and his staff had seemed pleased with the hotel and all of our work. I felt as if I were accomplishing something only a handful of advance people could: pull off a smooth hotel experience.

Staffers went quickly into their rooms. Some came back down the hall, probably headed for the bar. I kept waiting for Doug, or someone, to give me the all clear.

I felt fatigue hitting like bricks. Doug's room door was slightly ajar, a signal that I could knock and walk in and ask if they needed anything else.

"Stacy," said the valet, suddenly in front of the president's suite door only a few feet away. "He's asking for you."

I backed away from Doug's door. "Okay," I said, my eyes wide, my look saying, okay, I don't know why, but all right. The president had never asked for me before.

The valet slipped into the kitchen. I saw the president standing before me. Two million watts of power, light, and charisma were trained my way.

Behind me, the agent pulled the suite door shut.

"You want some cold water?" asked the president.

"Sure," I said. "Thank you."

"Would you like to sit down?" He looked toward the couch. I nodded. But I was stunned. Every fiber in me flared, seemed to say *no*, the order of things is that *I* go and get water and *you* sit on the couch. This was not about etiquette or chivalry. We operated under a clear system. He was the president, and I was staff. I did the tasks that supported his work.

But the president was getting me water.

I took a seat. Behind me, on a glass table, were the literary magazines I had rather boldly left in the suite. I had wanted to share with him what I was doing in my post–White House life. I worked for a nonprofit that placed writers in Detroit public schools. Three times a week, I went into local schools and taught creative writing to teenagers. I was so proud of their work, of my work, that I had brought their books all the way to Okinawa so he could read a few of their poems. We often left books or pictures in the suite for him to sign for people at his leisure. I did not think this was much different. Yet I don't know what I expected him to do. Headline a fund-raiser? Acknowledge the kids somehow? As if by his mere acknowledgment of our work, Inside Out Literary Arts Project and the students could benefit somehow.

I realized then that I was just like those other dreamers at the president's sleeves who hoped to alchemize his power into success for themselves. Yet I had only wanted to share. To enjoy his nurturing approval. Look what I've done since I've left, Mr. President.

The president was handing me a cold bottle of water.

The moments slowed and brightened like a videotape flooded with light. I still felt I could finesse the situation. If at any moment something went wrong, if this somehow turned sexual and anybody knew, this would be nuclear. And people always knew. The Starr Report had shown us that the light investigators and journalists shined on this man burned away any lies. Stories like these died in half-lives. Just ask Monica. No one wanted to go through what she had. To live with the shame of not just adultery but worldwide humiliation. To always know what others imagined when they saw you.

I liked the president, but I did not like him in that way. I did, however, feel attracted to his energy, as I think most of us did. Every staffer I knew wanted to be around him and felt special when given his attention. Why else would they devote so much of their lives to his success?

I also knew that staff like me were not asked to come inside the president's suite and sit down while the president fetched them a drink. In a year, I would see him again on another trip on another continent and introduce him to my fiancé. He'd say, "I feel like I've raised her," and talk about how young I'd been—eighteen— when I'd first worked for him in the White House. He could be so fatherly like that.

But in Okinawa the president stood behind the couch, next to the table where I had laid out the literary magazines. He opened a book filled with odes to mothers, to Belle Isle, to souped-up cars raced on Friday nights. He scanned a few. He looked impressed. I told him how much I enjoyed working with the students. His face lit up; he told me that this was the kind of work that changed lives. I told him I was writing, too.

I noted his silence. I wondered if that last statement morphed me into just another run-of-the-mill ex-staffer gone off to write about the White House. I had interned for three years for George Stephanopoulos, the commentator whom I heard called a "common traitor" for writing his book about his years in the White House, then joining ABC News as a broadcaster. I did not understand the vitriol of the Clinton loyalists who vilified George. Had it really been that terrible for George to write about his experiences in what I thought was a compassionate way, especially given what the president's scandals had put us through? Perhaps critics condemned him more because he dared declare, as early as January 1998, that impeachment was a real possibility. But so many still blamed him for his book, and I thought that their anger was misplaced.

I spoke again of how much I loved teaching.

The president looked up from the magazine and into my eyes. "You have a good heart," he said.

I was taken aback. That was not a light thing to say to another person. At least not to me. The words pierced my heart gently, swiftly.

"Let's go on the balcony," he said.

The balcony?

Not only did the president know the balconies were there, he was walking out onto them! If only my agent counterpart could see this. Would he laugh helplessly or turn in his badge? Alas, there was only so much you could do to protect the man. The president opened the glass and I followed him out onto the structure, into the inky blackness over the sea.

The thing is, you want this to be legitimate. You want there to be a real, beautiful reason that he's invited you here. Something about you. Good-hearted you. He's looking past the face and the body and sees something so deeply special about you. You want acknowledgment—all that comes when you've done a good job, when you're so *deserving*. You want that light. That hand on the shoulder. At least if you're like me and this sort of loving affirmation from authority figures still feeds you, even if you wish it would not.

On the balcony we stood, warm air around us and the ocean below. I was here, and he was there. He talked and I talked, but I didn't hear it all; I remember the scene like a series of night photographs taken with a powerful flash, trails dripping from the edges. He was saying he admired me, he'd always admired me. And I was staring back, on the brink of the greatest danger I'd ever known, and all I could see was the true deep black of the sky over the sea at night. The black that's gone in the morning.

The ship! Across the darkness I felt the Japanese ship protecting the hotel, the energy of a thousand eyes opening and focusing exclusively on us. *Click click click*. How do you say "Jesus Christ!" in Japanese? *Click click click*. I pointed the ship out to the president, but he didn't seem to care that someone must be getting a load of the fact that the president was on the balcony with a woman who was neither his daughter nor his wife. *Click click click*. I felt them trained on us. Hard.

The president stayed still. I was surprised. He knew we were being watched yet did nothing. The Secret Service did sweeps, and presumably they checked for listening devices in the suite, but I think the president believed that he wasn't thoroughly safe inside. That somehow, the balcony was best.

The men of the naval vessel had to be watching and filming and taking photographs. A technician must have been uploading and sending the images to Tokyo so the government could do what governments do: fatten the files. Somewhere there must be images of my frozen smile as the president said those banal things to me. His barrage of compliments as he told me what a great girl I was, making me want to believe him and escape him all at once. As each moment passed, my will to escape intensified, despite the numbness in all my limbs. An awful energy permeated my body. An energy of violation that was not yet violation. A coat of dirt that could be washed off but was not yet washed off.

He hugged me. Tight.

Click click click click click click.

I pulled away and said good night. Smiling. No sudden moves until I was off the balcony. I held my water bottle, and my heart pounded hard in my throat.

Have a good night. I have no idea what he said as I smiled; smiled and walked toward the suite entrance. I did not know how this would end. So far, no harsh words. No love words. Just me still pretending that what had happened was part of the normal course of business and that tomorrow there wouldn't be pictures on the front pages of newspapers with the question "Who is Stacy Parker, and why is she in the president's arms? Has Monica been replaced?"

His face betrayed no emotion. Was he irritated? Disappointed? I did not know. I turned toward the door.

Nothing had happened. Nothing beyond a hug that had gone a moment too long. But if people saw me leave the suite, would they see my flushed face and think otherwise?

No staff in the hallway, thank God. No Chelsea. Only the posted agent, and he looked unconcerned. Maybe this is what it's like to have a bullet whiz by you and not get hit. It could be so hard being a woman around President Clinton. Get into a moment like this, and you felt powerless. He was the president, so you were the problem, not he.

I made it back to the atrium railing. Safe. I stared at the quiet lobby below, with its tiles and indoor plants with shiny leaves. I felt sick to my stomach. Nothing had happened, but something could have. I could have made a mistake just then. One mistake, and suddenly you are the second Monica. The line between safety and doom was so thin when the president stared you in the eyes.

KEEP IT TO YOURSELF. Tell the story, and you become Delilah. You become the woman bent on undermining the great leader and ruining the good thing for his supporters and staff.

But that's politics, no? People get killed, and you learn where the bodies are buried. Sometimes you're the witness. Sometimes you're the killer. And sometimes you're the person who gets shot in the skull. I mean this metaphorically, but I left political life because I did not want a life of secrets, fear, and submission to power. I knew that what I ached for was love. The love I felt working in a classroom, when I instructed and nurtured my students well, exercising a power more interested in guidance than in control. Or coercion.

The day after the balcony scene, I walked President Clinton to an event with then–British prime minister Tony Blair in the hotel. He acted as if nothing had happened. No one pulled me aside and said that from now on I was unwelcome on the road; in fact, no one else seemed to know. Later that evening, our director of advance, Brian Alcorn, asked me if I wanted to do a RON on the upcoming Africa trip. I said yes, thrilled to be given the opportunity to travel and experience the world beyond my Detroit origins, with airfare, lodging, and per diem paid for by the government. And on that trip, that very next trip, I would meet the man who would become my husband, Kai Aab. As a young girl, I would never have guessed that there would be so many pitfalls to leap in order to find the love I desired.

Everything had begun so differently. As young women in the Clinton White House, we interns and staffers shined. We believed. We worked like crazy. And except for one exceptional woman, we kept our mouths shut—unless we were appointed spokeswomen or instructed otherwise by interrogators. That's what we were supposed to do: Be a team. Be loyal. Climb even higher. But you had to be careful. Make mistakes with a powerful man and they would turn your name into mud. Just as they did to her.

I

INTERN

2

Entrée

Washington, D.C.
1993

I FEEL LIKE I'VE RAISED YOU. THE PRESIDENT COULD SAY THAT TO ME because I began White House work at the age of eighteen. If I wasn't the youngest volunteer in the complex, I was close, back when I assisted with "news analysis" duties in the Office of Media Affairs in spring 1993. Volunteers summarized the evening news broadcasts for inclusion in the president's daily clips—an inch-thick stack of articles from the *New York Times*, the *Washington Post*, the *Wall Street Journal*, *USA Today*, the *Washington Times*, the *Boston Globe*, the *Los Angeles Times*, and other papers, cut out and photocopied by an overnight staffer who had them ready for delivery to the president and his top advisers by sunup. News mattered in the White House, and I felt honored to contribute to that critical operation. My efforts were recognized that May, when my boss, Julie Oppenheimer, took me to the West Wing to be interviewed for the position of George Stephanopoulos's lead summer intern.

George Stephanopoulos, the communications director to the president. Or, as we knew him then, the young, gorgeous man at the podium we all ached to glimpse in person. If any of my fellow Democrats in our freshman hall had seen him out in the world, calls would immediately have been placed to our friends at Big Ten schools. *Sure, you've got Michigan football, but guess who I just saw? George Stephanopoulos!*

Girls like me loved everything about George. We had come to Washington, D.C., to study politics, so he was our role model. We admired his brains and his meteoric rise, yes. But we loved his person, too. His hair. His glasses. His calmness facing down the daily press storm. We loved him even if he wasn't tall. Even if jealous boys down the hall said he was gay. (*No way!*) He was our crush. He was who we wanted to be, or to date, by the time we were thirty-one, that young age at which he had first achieved his acclaim as the president's adviser and spokesman.

Going to college in D.C. made it easy to follow in our heroes' footsteps—literally. Just months into my new life as a George Washington University (GWU) freshman, I joined the Clinton effort as a volunteer for the 1992 Presidential Inaugural Committee (PIC), whose offices were in the GWU student center in Foggy Bottom. As a result of walking into the PIC offices to volunteer and then working hard, I was offered the chance to work in the White House communications office—overseen by George himself. That spring, I worked three afternoons a week in the Office of Media Affairs, housed in the behemoth gray Old Executive Office Building (OEOB) that bookended the White House to the west, as the Treasury Department bookended it to the east. When George's executive assistant, Heather Beckel, told Julie that they needed help in their personal office, Julie recommended me for the job. Julie delivered this heavenly news to me as we stood on a stone balcony overlooking the White House's North Lawn. Below stood CNN's Wolf Blitzer doing a live shot, washed in hot white camera light. If we walked inside the office, we could have seen the tight shot of his face, back when you saw only the correspondent speaking and not a constant news crawl across his chest. But we remained outside. Julie explained that as George's lead intern I would really be Heather Beckel's lead intern. "She's been with George since the early days in Little Rock," said Julie. "She's tough, moody, but fair." Julie said that I would serve as Heather's liaison with the other summer interns.

I later learned that kids from the official intern program tended to be rising juniors and seniors from impressive East Coast schools, many of them Ivy League. Some had "FOB" connections—family who were "Friends of Bill." Their sessions lasted half the summer. I, on the other hand, could begin immediately and stay indefinitely—the benefit of going to college three blocks away—without having to apply to the program. And if Heather willed it, I would remain a "volunteer" but would still be their boss.

If you're good, come on, let's go—that's what seemed to matter in the White House. I loved this about the Clinton administration: Work ethic mattered. Performance mattered. And if we opened *Time* or *Newsweek*, we knew our performance was scrutinized. Hard. The administration had already been rocked by two failed nominations for attorney general. The president had recently announced his "Don't ask, don't tell" policy, his attempt to mollify supporters and opponents of gays in the military that managed to please no one. Commentators questioned the president's commitment to core principles. Journalists sniped that the White House was run by "kids" who didn't know yet how to run the place. Yet what I saw were rooms full of men and women devoted to major policy initiatives, who typed, worked the phones, and stayed at their desks late into the night trying to improve the lives of everyday Americans. How honored I felt to be asked to join them. Only a year before, I had been running student government meetings devoted to small-scale service projects and homecoming floats. Now I was contributing to the fight for real structural reform in Washington. Down those OEOB halls of diagonal checked tiles I walked, giddy sometimes, knowing that the future looked bright, especially for a young black girl like me, who, half a century before, would have been seen as suitable only for household help, no matter my smarts or experience.

To be embraced by like-minded comrades in government, this was my dream. Opposite to the nightmare I had experienced two years before, the day I had received the scholarship letter from the Central Intelligence Agency. A student office aide had entered my Physics II classroom, giving my teacher, Mr. Woodside, a sealed envelope.

"Eleanor Parker?"

Mr. Woodside called out my dreaded first name, signaling that whoever had written me had found me through an official list, for anyone who knew me personally would know that I went by my middle name, Stacy. He glanced at the sender, and his eyebrows rose. I stood up and retrieved the letter.

The CIA!

Mr. Woodside began writing on the board, and under cover of that light noise, I opened the letter, not believing that this most mysterious of agencies would contact me at school. This was the stuff of TV spy shows aimed for kids.

"Dear Eleanor Parker, Because you are an Outstanding Negro Scholar . . ."

Ah, affirmative action at work. Not bad, I thought, appreciating their efforts to diversify their ranks. If I were accepted into their program, a full college scholarship would be mine, as long as I committed to summer internships and a few years of service with the agency after graduation.

College tuition was an issue. My parents had said they'd pay for my university schooling as long as it did not cost more than the University of Michigan—an offer that had seemed generous at first hearing but ruled out the options I cared about, like private schools such as The George Washington University, which charged double to three times the amount of a local public university. If I were to apply to the CIA program, suddenly the worries would disappear.

I sat in my metal chair with the welded-on desktop, students behind me, students in front of me, and thought *whoa*, the CIA and—

Me.

At Troy High School, in suburban Detroit, I debated on the debate team, role-played diplomatic crises in Model United Nations, and served as class vice president. Senior year, I had led as student government president. Yet somehow, growing up a black girl in the suburbs, I had never felt like an insider. Memories of elementary school ostracization died hard, and when I hit middle school I lost all interest in assimilation. I wore black all the time. I shaved off some of my hair and teased up the rest, just like Robert Smith of The Cure. I wore black leathers and buckles, and though I was never a full-on punk, I was punk sympathetic and found myself in constant protest against the status quo. My parents never made political contributions, but I did—my first was to Jello Biafra's "No More Censorship Defense Fund." At age twelve, I had mailed away ten dollars of my babysitting money to defend Biafra's band, the Dead Kennedys, from having their albums blackballed from record stores because of an H. R. Giger art insert deemed pornographic. Jello Biafra sang boldly of public hypocrisy, and I wanted to be like him: fearless, telling the truth as he saw it, for I idolized musical outsiders who knew how to be heard, who made it so you wanted to hear them.

In middle school, my friend Susan and I wrote out Dead Kennedys song lyrics on our paper-bag book covers. Songs such as "Holiday in Cambodia," "Moral Majority," and "Rambozo the Clown." Jello Biafra and his band articulated responses to government deceit and the narcissistic money love of 1980s America. I ate their music up, needing to hear people speak the truth as I understood it, not as so many authority figures tried to shape it. I was no expert on

CIA actions abroad, but I knew about the Iran-contra scandal and the history of U.S. attempts to overthrow governments all over the world. What analysts called blowback, the often unintended and unexpected consequences of this murderous meddling, I understood as something much simpler: karma. To every action there is an equal and opposite reaction.

And they wanted to hire me?

Well, maybe the CIA had vision. Others had mistaken my black features for those of a foreigner before. Spy shows tended to depict spy women as white and in need of wigs and makeup to slip into native disguise abroad. Not me. I could skate around the tightly wound cities of the world with their gold-leaf domes and look as if I were from the place. Guards and gunmen would look at me and think, yes, she's kin.

But then I imagined a helicopter piloted by Iran-contra mastermind Oliver North. There's no more space in the copter and there is fire all around in a Central American jungle and there's me, the operative, on the ground. He sizes me up. *Dispensable.* And just like that, he flies to safety and I'm left to die. All that devotion to mission, just to be abandoned. This was my deep fear, this just desert for trusting these men of power in the first place, men who never had my interests at heart.

No use prolonging the decision making, I thought. I tore the letter into two, then four, then eight pieces. When the bell rang, I dropped the shreds into the garbage bin beneath the pencil sharpener.

A few months later I applied for a full scholarship from The George Washington University. Because I was an *Outstanding Negro Scholar* with so many extracurricular activities, they gave it to me. No reconnaissance duties necessary.

THE INTERVIEW DAY ARRIVED. Step onto 1900 F Street NW, turn right, and there it was: the Old Executive Office Building. My roommate, Elyse, hugged me and wished me good luck as I departed and quickly walked the three blocks toward Seventeenth Street. My street felt like an asphalt valley between the mid-rise office buildings fused together and sliced off at the top, their heights not allowed to be greater than that of the Capitol dome. I entered the Seventeenth and G Street entrance grinning from ear to ear, knowing that my name was in the system. That *I* was in the system. That armed guards would step aside to let me pass.

They did. Once inside, I slipped over to the ground-floor bathroom. I stood before the full-length mirror and used my fingers to comb through my hair still shaved up a few inches from the nape of my neck, hair that fell softer now that I had stopped teasing it and squirting Mink hairspray all over it. I smoothed down my gray suit with the knee-length skirt and checked that my flats remained unscuffed.

Thank God I had the clothes. Suits. Shoes. A drawerful of hose. My mother took me shopping whenever I went home. She wasn't prissy or pretentious about it; she just wanted me to be prepared. Lord & Taylor at Lakeside Mall. Or Hudson's at Oakland Mall, just down I-75. She even let me keep her Saks card. I tried not to use it—my mother didn't make that much money as a nurse—but the card gave me security, in case I had an event emergency and needed something quickly.

"You ready?" Julie asked with a smile, rising from her chair as I entered her OEOB office.

I idolized Julie. I loved her tomboyish look and pretty smile. I was in awe of her background, too: she had worked for the "gonzo" journalist Hunter S. Thompson. I'd read *Fear and Loathing in Las Vegas* as well as some of his 1992 campaign coverage. Julie was such a cool girl, I imagined her fitting in perfectly with that star of the counterculture.

"I'm ready," I said.

"You look nice," Julie said. "Come on, so we're not late."

Down the stone stairs we went and across West Executive Drive, the asphalt moat that divided the OEOB from the White House with its walls like smooth white frosting.

I always expected the White House to rise mighty like a mountain, but the complex's wedding cake structures were modest compared to what the corporate equivalent would be in Manhattan or Silicon Valley. Still, they commanded respect and awe.

"Don't be nervous," said Julie. "You're going to do fine."

We walked through the ground-floor entrance of the West Wing. On the entryway wall was a poster-sized portrait of President Clinton in a narrow wooden frame. Ahead sat a big-chested, black Secret Service agent who wore a coat and tie. I felt my breath stop. Guards could make me feel criminal until proven otherwise.

"She's with me," Julie said, barely slowing her stride.

Julie's blue badge gave her instant access to the grounds and proved that she belonged there in the West Wing. She hooked left and took us down a short hallway. She made a 180 and walked up a narrow flight of stairs on cushy blue carpet, making me remember back to when I was ten and our family had picked out carpeting for the family room, living room, and bedrooms of our new Troy home. We had bought nice padding then, but our carpets had never felt like this.

"You're going to see something cool now," Julie said in a hushed voice. We exited the narrow stairwell replete with more wood-framed shots taken at presidential events and entered the first floor of the West Wing. We turned left and walked down a narrow hallway, passing another open door—inside, a huge color portrait of Earth as viewed by the astronauts from space. "That's the vice president's office," Julie said, sotto voce. "That's the chief of staff's office," Julie said as she motioned straight ahead, waving hello to young women sitting behind dark polished desks as we hooked another left.

She pointed her gaze at an open door ahead. "That means he's not here," Julie said as we approached a short rope between two stanchions. After walking past two more open offices, with women inside answering phones, I glimpsed the telltale concavity, and I knew that yes, that was it. The Oval Office.

But Julie kept walking. I wanted to stop and gape and take it all in, but we were not tourists today. I had an appointment to keep.

"This is Upper Press," said Julie. "Down there is Lower Press," she added, pointing down toward another office connected to the White House briefing room. Beyond was the cubicle honeycomb where the White House press corps wrote and filed their stories—a sight that shocked me later when I first entered it, for I had never imagined that men and women at the top of their profession would be assigned to such cramped quarters.

George and Heather worked in Upper Press, since George was the director of communications. He also gave the daily briefings and had the large press secretary's office that looked out over the North Lawn. We entered the Upper Press anteroom, busy with assistants on their phones. I scanned the desks, wondering which of these whispering women would be my next boss.

Julie stopped at the last office door. She looked inside. "Heather?" She looked back and motioned me to follow.

Wow, I thought, what a big, sunny office this is, with TVs up high on shelves, muted and tuned to different channels. Out the windows, you could see the white columns of the West Wing's entry portico and the long driveway headed to Pennsylvania Avenue where the White House correspondents gave their stand-ups. A strikingly beautiful young woman walked toward us. Black straight hair and black straight bangs. Pale skin. Light eyes. She reached out her hand to me.

"You must be Stacy," she said with an English accent. "I'm Heather. Pleased to meet you."

I smiled and gave her a firm handshake. Heather had been alone in George's office, and we followed her out to her nearby desk in the anteroom. She pulled up a chair and asked me to sit. She greeted Julie, and after a few words, Julie left the two of us alone.

I wasn't listening so much as taking her in. She kept her hair in a longish black bob, precision cut. Pretty, definitely pretty, I thought, with blue eyes and the kind of voice that could make someone react, either to her personal power or just in Pavlovian response to the sound of English authority. She wore a long jacket, a straight skirt, boots with heels. She wasn't tall; even with the heels she came up only to my shoulders, max. In a profile, I'd later learn that she had been born in Arkansas but raised in England: the journalist called her accent "clipped." To me, she sounded like a beloved girls' school teacher from deep in the posh countryside. The kind of teacher who broke hearts, including those of the girls in her charge.

"I need you to see something," Heather said. She pointed to two big white U.S. mail bins filled to the brim. "There's more, much more. This is why I need you. I am only one person. I can't handle this by myself."

She explained that their office was drowning in mail. Fan mail. Issue-driven mail. Mail from schools and churches that wanted George Stephanopoulos to speak at their events. Mail from love-sick girls who wrote fawning prose. Some wanted pictures ("yes, we have pictures"), some wanted dates (Heather's *you have to be crazy* look). Regardless, the writers deserved responses. But writing them required time and energy—and, by the looks of the bins, an army of laborers. Heather's solution was to get a supply of summer interns detailed to the correspondence operation, secure a room in the OEOB, and have one intern take charge and report directly to her.

"Does this sound interesting to you?" she asked. "Could you be my right-hand woman?"

"Yes!" I said. A smile spread across my face so wide it hurt.

"Then you will be," she said.

Heather added that she and George were changing offices. Good-bye to the (sunny, spacious) one in Upper Press and off to the new one, the only office with a direct connection to the Oval Office. I had heard rumors that George was facing demotion, that he wasn't handling the press well enough during this challenging first year of the administration. Heather read my mind and said, "If you want to call not having to deal with those animals every day and now being a senior adviser to the president with an office connected to the Oval a demotion, then you may." Regardless, former White House adviser David Gergen would soon join the staff as counselor to the president. The "kids" would now have to share the wheel with someone who could be clearly seen as an elder, one of the graybeards who George would later admit the incoming staff had thought they did not need.

Then right into Upper Press walked George. He had a tired look on his face—but that did not matter. George R. Stephanopoulos did not disappoint! Yes, he had that handsome way that made the smart girls swoon. No, he wasn't tall, but so what?

"George," Heather said, "this is Stacy Parker. She's going to lead our correspondence effort." I stood up. George shook my hand; a nice, firm grip as he met my eyes. "Thank you," he said, looking genuinely grateful for the help.

He did not linger. I felt honored that he had stopped at all, knowing how busy he was as Heather handed him his full call sheet and he took it without comment. I knew that, at that moment, I was no different from any of those girls whose letters were in the mail bin, who watched him on TV and saw and felt so much possibility for themselves, and our generation, when he spoke. He entered his office and closed the door.

"So that was George," Heather said, taking in my happiness. "Now we're going to need to get you a blue badge, because sometimes you'll be working over here," she said. Could she hear my hopscotching heart? "That involves an FBI background check. I'll see what needs to happen. I am really grateful you're on board with us."

Working in the West Wing? An FBI check? So much to process and no time to do so as Heather walked me to the lobby, telling me when to report back. I thanked her, trying to tamp down my enthusiasm so I didn't look too much like the kid I was. A kid who, at eighteen, was now Heather Beckel's deputy.

WHITE BINS OVERFLOWING and heavy with letters. Each with the residue of hands, machines, and the miles traveled just to rest en masse at Heather's feet. Small mountains of need waiting for her determination of what would be opened now and what would wait until assistants could do the sifting and responding. I knew that congressional interns spent most of their time answering phones and constituent correspondence, so at no time did it cross my mind that this job could be beneath us. The previous summer I had been thrilled to be stuffing envelopes for the Oakland County Democratic Party. The older ladies had laughed at my earnestness, but I didn't care; I was doing my part for the cause. Now I was being tasked to help one of the shining stars of the administration get a grip on his correspondence? What unbelievable good fortune. I called my family and my friends Christina and Peggy back home and told all of my friends in the dorm. Hearty congratulations came from all.

I RETURNED THE NEXT DAY as instructed. I entered the ground-floor entrance of the West Wing, telling the agent that I was there to see Heather. He pointed me to the couch as he placed the call for me. While we waited, I learned his name was Brent. He spoke with a low, sweet voice. He made me feel welcome. Military personnel passed by briskly, upright, like thick cords of wood bound together. They moved purposely toward what I assumed was the Situation Room, and I felt a sensation that would become very familiar: the contact anxiety of other people's worry.

Heather rounded the corner, looking as serious as the military men. When she saw me, she smiled. "Let's go see the office in the OEOB," she said, and I followed her back out into the sunshine, back up the stone stairs.

If you spoke above a whisper in the OEOB, you heard the echoey dissolution of your voice in those cavernous halls. Yet I heard her instructions crisply as we click-clacked across the black-and-white diamond floors. Heather wanted to have every letter responded to by the end of the summer. She wanted us to create several templates that we would personalize as necessary. She would provide samples of past responses, and we could shape and evolve our replies.

Then she repeated the magic words: "You'll be working half days in the West Wing. I need your help with the phones and with special correspondence." So there it was: no second thoughts, no forgotten offers. I knew that only a special few of each semester's official interns worked in the West Wing and that most were assigned to offices in the OEOB. I might not have been in the official intern program, but my opportunity had come gilded because Heather believed I could do a good job.

We rode the elevator and made our way to a room tucked away in the far corner of the fourth floor. She stuck the key in the door, and *voilà*.

The room felt as if it had been locked since the Bush administration. It was a cubbyhole, really, with a red couch with thin, fabric-covered cushions. A few heavy desks. A computer. A connecting room with a table and phone. Walls that needed fresh paint. The West Wing this was not, and that was fine. This was ours now. Mine. Anchored by the three full bins of mail on the desk. I noticed that the heavy wooden door had a lock. Hmm, I thought, a room with a couch and a lock. I wondered, smiling to myself, what the Bushies might have done here.

"I could come here and sleep," Heather said longingly. Suddenly I was in her shoes, feeling in my bones the fatigue that seeped from her voice. These guys hadn't had a break since 1991. The campaign had been a marathon with a full sprint at the end. If you won, there was no break: you needed to lay the groundwork for those first hundred days, all the while jousting for who got what in the administration. Then boom, you ruled the world—work that seemed no easier on a staffer's constitution than being a galley slave on a ship.

Heather stared at the bins of mail. Then she turned to me.

"You still want to work with us?"

I sensed a pleading in her eyes. For the first time it occurred to me that she wasn't the only woman with power in that room.

"Yes," I said, smiling.

But as we locked the door behind us, I felt the first fingers of dread, ghostly, slipping in through my kidneys to my stomach. Those stacks were like hay that needed to be spun into gold. No firstborn child was at stake, but my reputation was, and all my possibilities in politics, because, having been handed this assignment at eighteen, I believed that my whole future in government rested on how well I performed—for what else did these people have to judge me on? Just my performance. There could be no falling back on family connections. There were none.

I knew instinctively that pleasing George was beside the point. Heather was who mattered. I watched her walk back across West Exec. So much rested on the shoulders of that petite body, and though I thought she was pretty, I sensed one thing: she was a tank. She knew how to get what she wanted. I walked back to the elevator and to my waiting piles of mail, grateful that what she wanted was my help. I had no intention of letting that woman down—ever.

The pressure I felt, at that moment, made me wince.

3

Paradise City

MAY 1993, AND WE WERE GOING TO PROVIDE EVERYONE WITH health care and fix the economy. George had just begun his new role as "senior adviser for policy and strategy" and I was his new intern, fresh to the West Wing ecosystem, figuring it all out.

At school, I learned that presidents passed their proposals by negotiating with lawmakers and using the "bully pulpit" to appeal directly to the public. At work, I watched George ride herd over these efforts. He now lived a life of constant prioritizing, debating, and decision making, managing the nonstop machinations necessary when you wanted to sell the nation on your agenda. George had advised the president before, when he was director of communications. But now Dee Dee Myers gave the press briefings. George could focus on strategizing full-time, while working the press one-on-one; journalists still burned his phone lines looking for quotes and explanations.

With the new job came a new office. George's new space was small but geographically desirable: one door led directly to the Oval Office via the president's study and pantry. No other senior staff offices enjoyed such intimacy. His main door led to a small anteroom, a bufferlike space between his office and the hallway.

Heather Beckel sat in this anteroom, the pale walls creating a tall, tight box. Her white desk jutted out from a credenza of white wood. White cabinets were mounted above her head, and another framed poster-sized portrait of President Clinton hung high behind her back. I think of this place as so white with light, the air electric yet encased by the smallness of the room, like the light of a bulb.

The problem was that whoever had designed this space had created it for a single secretary, back when communications were typed on manual typewriters and phones had single lines. By 1993, the volume of calls, letters, and now e-mails was too much for a single secretary. An intern sat between Heather and the door, wedged into the corner by the in-boxes and the second computer.

Often, that intern was me.

"Heather, do we respond to people who write George asking for the president to come to their event, or do we send the letter straight to presidential scheduling?" I asked her this on my second day. Heather seemed to be free to answer a question, for she was just reading something on her computer screen.

"Heather?" I repeated. "Heather?"

"Stacy, did you check through the sample letters I gave you?" she asked.

"No, I'm sorry," I said. "I didn't check."

On the first day she had given me a stack of letters to respond to and a folder of sample letters to use as models. She issued a few other quick instructions, a thank-you, then returned to the thousands of tasks that always seemed to clamor for her attention. *Get to work,* I felt her say, even though she did not say it. *Be useful. Now. Fast. This is how we respond to requests. This is how we prioritize. Interrupt me for this, but don't interrupt George for that. Lower your voice. Lower. Don't be lazy—look everywhere you can for answers. Did you research? Did you think before asking me? You work for me, not the other way around. You're here to be value added. To me.*

As I allowed Heather's unspoken rules to pour into me, I thought of the girl in Jamaica Kincaid's short story "Girl," the one who received instruction after instruction from her mother. I went to the folder and found a previous letter that had dealt with presidential scheduling requests. "Look with your eyes, not with your mouth," my stepfather used to say, and I resumed work, even more diligently than before, wanting her to know that I got it, that I knew I needed to be an assistant who didn't suck up her time with long explanations of how things worked.

That was the catch-22, of course. So many tasks involved personal preferences that were neither intuitive nor spelled out, from lunch orders to which callers trumped others for George's attention. The president's close friend and power broker Vernon Jordan? Interrupt him. Ann Devroy, the *Washington Post*? Go check. Same for the *New York Times*, the *Wall Street Journal*, CNN, or any of the networks. Callers who claim to be friends of the family? Just take a message.

I watched George and Heather work so hard to "put people first," as the Clinton presidential campaign had promised, that nothing Heather asked seemed unreasonable. I felt like a teammate now, believing that my contributions to George's office would help please constituents and solidify support for President Clinton's agenda.

MOST OF THE OTHER CORRESPONDENCE interns were male and older. Heather started with three, and soon we expanded to six, including me. I wasn't intimidated by the boys. Not at first. Freshly deputized and imbued with the sense that we had a crucial mission, I plowed ahead into this unmapped territory of management.

Had I supervised other people before? No. But that fact seemed immaterial.

We brainstormed together in the 415 OEOB cubbyhole office. What were we going to do with this mess? I talked with the other interns, including Doug, another GWU student volunteer who had started in the spring. We came up with a plan. We would open and read the letters, then code them by issue (for example, health care reform) or request (autographed picture). Draft specialized letters of response. Then mail merge: stick in the name and the address, and presto, automation. If some letters struck us as poignant or eloquent or somehow deserving of more personalization, we would go back and stick in a line that was specific to the letter—but not in all of them, so we could be fast. If we kept up a good pace, we should be able to hit that efficiency sweet spot: the greatest amount of input with just enough personalization required to make recipients feel heard.

Initials ____   Date _____

TYPE:

___ Invitation   ___ Forwarding to: _____

___ Gift: _____   _____

_____   ___ Repeat

___ Legal   ___ Wants Picture of:

___ Suggestions   Autograph also?   yes   no

___ Specific Suggestions: ___ What did you do as Dir Comm?

_____   ___ Class letters

___ Can I Write Suggestions?   ___ Students

___ Support   ___ Kids

___ Complaint   ___ Flattering letter

___ Resume

___ Internship

___ Article

___ Gays in the Military

HOLIDAYS:

___ Birthday Card

___ Nameday Card

___ Valentine's Card

___ Christmas

RESPONSES WILL BE:

___ Late

___ Prompt

SPECIAL:

___ Wants G.S. Advice

___ Wants FLOTUS or POTUS
to come to an event

___ Other: _____

_____   ___ **FOREIGN**

We nailed down our course of action. I shared our plan with Heather. She nodded and smiled. "Well done," she said. "I knew you could do this."

From that day forward, I coded letters and managed the operation, and I jumped into both roles with gusto. But I learned a crucial new fact about myself: when under pressure to deliver for people I admire, I wanted perfection, and I could charge ahead like a Heisman Trophy contender with no regard for whose feet or feelings got trampled beneath my cleats.

This doesn't sound right. / Can't you add more here? / Here are revisions. / No, I'm not giving it to Heather like this. / (Yes, you can look at me like that all you want, but you still have to redo it.) / Can we get this all done by 5:00 p.m.? / Seriously. / I mean 5:00 p.m. / (Why does this feel like pushing a full crate uphill?) / Why is he asking me to do work that he can do himself? / I need you to write a memo outlining what 415 did this week. / Yes, I'm giving it to Heather. / Let me help you. / Let me do a draft. / I'm sorry, this just won't work. / This letter is from George Stephanopoulos! / This doesn't read well enough. / Please, will you make these corrections?

If you could only have heard the urgency in my voice, both out loud and in my interior monologue!

In my own defense, this was the beginning of a peculiar way of living for me: the sense that every action, no matter how small, could make or break not just my career but my future. No one ever articulated this threat. Yet I felt so seized with expectations that there was no experienced difference between getting the coffee right and the letters done well and performing lifesaving CPR, had I been called on to give it.

A pattern emerged: Fear. Performance. Evaluation. Fear. Performance. Evaluation.

And sometimes, pleasure. "Good job," Heather would say. Sometimes George would say it himself. Later, when he wrote that I was "going to go far" on an autographed picture of him standing with me and my family outside the West Wing lobby, my fate felt ordained.

This was the promise of power, no? Work hard for the principal and for your bosses and you'll be rewarded. I threw myself into my duties. Their needs were now my own.

WHAT'S GEORGE LIKE? Once I was working shifts in the West Wing office, I was asked that so much. Let me say this: He was cool, steady. He drove a red Honda CRX with a Clinton sticker on it. He carried a Bible in his backseat. He lived off Dupont Circle, my favorite part of town, to which I often gravitated with my gay friends, in the skinny building with the new Starbucks—*yeah, look for the car, you'll see it there.*

In his 1999 book *All Too Human*, George wrote that he had "flattened his feelings" to deal with the daily bombardments of his work life. I had no idea what his interior life was like. He just seemed unflappable to me. In the book, he revealed that he had grown a short beard to cover stress-induced hives. All I saw was the beard. People in interviews called him pessimistic, but I never felt that either. The thing was, he never showed much emotion. He never screamed or sighed or slammed the phone down. He never raised his voice.

George was not like his fellow adviser Rahm Emanuel, who practically ran into George's anteoffice sometimes, half dancer, half storm trooper, stopping on a dime and talking to one of us, usually the one who was not on the phone, waiting for George to get off his phone and deal with the crisis of the moment. Rahm gabbed. Rahm flirted. George was not like that. He kept a distance. Respectful. But not cold: he never walked in with ice in his eyes.

I remember when I first met Harold Ickes, the deputy chief of staff and Democratic party warrior who was contributing to Health Care Task Force efforts but also running point on Whitewater, the burgeoning scandal around the first couple's previous business dealings. I once complimented him on his floral tie. He looked at me strangely, for I suppose I was the only one who didn't know that this was the tie he wore every day. At least every other day. But I soon learned that as a civil rights activist in Louisiana in the 1960s, Harold was once beaten so badly he lost a kidney, and that his father, as a member of Franklin D. Roosevelt's cabinet, had desegregated U.S. government facilities. I would not say anything bad about Harold Ickes ever again. Even though he could be rather cool.

But George, he kept to himself. At least with me and the other interns and later, the junior assistants. And we protected him with a quiet fierceness. That was our job. No one told us that that was our job. It just seemed natural to do.

WORKING, I WANTED THAT FEELING of rowing on the Potomac River, that feeling in the eight with all of us pulling our oars. Sixteen arms and sixteen legs powering that slim boat onward, as we were led by our coxswain, as our coach called out to us from his motorized boat nearby. The freshman women's crew would row from Three Sisters to the Fourteenth Street bridge, and yes our muscles ached and our mouths dried up and our hands grew ruddy and callused, but we all needed one another to succeed. One fumbled stroke—stop catching those crabs!—and we all felt the stumble, like landing a foot through a rotted plank. No one wanted this. We paid attention to one another, synchronizing our bodies. I'd never known teamwork like that before. All my sports before had been individual events. Eight hundred meters. Shot put. Discus. Though I'd had a debate partner before and had competed on Model United Nations teams, I'd never known the daily pleasure of working together with peers.

Yet I quit crew after freshman year. I liked my freshman crew teammates, but I was unsure about the varsity rowers. The previous Christmas break, all the rowers had headed to Tampa, Florida, for a one-week rowing camp. Two hard practices a day. Eating every meal together and bunking in gymnasium-like accommodations, and suddenly we were together as never before. I tried hard to be friendly and open, but I remembered how some of the older girls had snickered about my boots when I had worn them with jean shorts in the evening. Or they made fun of my haircut for its asymmetry and fluffy wave. I felt as if I'd been air-dropped back in time to middle school and there was no escape—I couldn't go home or hide inside my headphones. Crew camp reminded me of how it hurt when the group turned against you. When a majority decided that you were the butt of the joke.

The boys hurt me worse. I had a crush on one varsity rower. He noticed, and he began talking to me. Later that week, we took walks together. We even walked to the University of Tampa football field and kissed on the fifty-yard line. But other boys stared at me coldly. It wasn't just lack of interest but contempt, somehow, in the eyes of some white boys, who found me sexually undesirable. There was something mean in their eyes that I couldn't figure out. When his teammates learned that we had hooked up, he was suddenly on the spot. "Do what you have to do," said one male teammate. *Do what you have to do?* It was as if he had to be excused for choosing me to kiss of all the girls present, freshmen and varsity, so many of them ostensibly preferable since they were long, lean, and white.

I credit the daily 6:00 a.m. practices with instilling the discipline that allowed me to flourish academically and with making me so fit. But at season's end, I hugged my freshmen teammates and my coach, Jack, thanked them, and told them that my schedule was now too full for crew. My friends I'd still see when I wanted, but as for the others—including the rower I kissed—that was the beauty of an urban campus: it was so easy to ignore people who hurt you, for there was so much more to our lives than our cliques, as there was to mine that summer I began at the White House.

WAS THERE A BACKLASH to my pushing of the other interns? I don't know. They performed. I performed. Heather complimented the work, and I passed on every compliment that Heather did not give them directly. Doug and I had GWU in common and got on okay, but I remember my awkwardness with most of the intern program boys, with their crisp white shirts and expensive blue ties—boys who looked to me as if they had all pledged the same fraternity. Sometimes I entered 415 OEOB and noticed that all conversation stopped once I opened the door, and did not resume.

I wondered if they had been speaking about me. Gossiping about how I acted as if the sky might fall if we did not finish our self-imposed quotas. Surely not—those boys had better things to talk about. But the fact was, my managerial role kept me separated from their group. I tried not to let this bother me, but sometimes it did. Sometimes I wanted to be part of the group, even though they reminded me of the varsity crew boys sometimes. That made it harder for me to be brave and try to be friends.

I focused on keeping Heather happy. It was a smart plan, I felt, since she was our boss and the ultimate judge of how successful we were at our tasks.

"FOLLOW ME, STACY," Heather said to me one June morning as I walked in the West Wing office door. I put down my bag and let her lead me past the Oval Office and the smiling Secret Service agents acknowledging us as we passed, through the glass doors out onto the Colonnade.

The Rose Garden! I saw an audience sitting in rows of chairs. "The president will be right out," Heather said. "Take a seat and enjoy yourself. And don't worry," she added, "I have the phones covered."

I sat down, my flats sinking into the thick green sod, the tall southern shrubbery around us so much thicker and lusher than what I had known growing up in Michigan. You would think scent rolled off everything like wafts of night-blooming jasmine, but the thickening air seemed to keep molecules buzzing yet steady. Nature felt like a thousand muffled amplifiers as we waited for the president to enter.

Wait, who was that? I spied a gorgeous young man. He looked at me, too.

I took a Polaroid in my mind of me and that boy. This is how we met, we'd say, smiling, still holding hands, our grandkids embarrassed by our affection. We met on an early summer morning, with the heat still gathering and the grass around us damp enough to wet our clothes if we rolled around in it, kissing, or just tickling each other—

President Clinton entered the garden and we rose, clapping, then sat again as he began his announcement from the podium. How impressive that he was so close, with that gray hair going white, the energy radiating from his body so strongly. The substance of his speech I've long since forgotten, but the boy and I both sat at attention in our business suits. I listened, humbled to be there, but I also sneaked glances at the boy behind me, whose hair glinted in the sunlight.

A president's daughter, Tricia Nixon, was married on these grounds. Later, President Clinton's nephew would also be. I'd never been the girl to buy wedding magazines and dream of ceremony, but

I knew this was the fantasy setting for a public union, for a reception with a thousand guests spilling out onto the South Lawn as if it were the Fourth of July.

The event finished. I could have been like the others, the congressmen, guests, staffers, and interns looking for a way to snare the president's attention—indirectly, of course. *Never, ever appear to be a groupie of any kind.* Instead, I stood near a column, calculating that I had at least a few minutes before I needed to call Heather to retrieve me.

The young man walked my way. He introduced himself. *Charles.** He asked my name. *Stacy.* He had curious eyes, yet kept himself at a polite distance. He asked me what had brought me there that morning. I told him I interned for George. He sounded interested in my work, but he looked even more interested in me, the girl.

Then he did it—he asked me to lunch. I said yes and gave him my phone number.

The president kept shaking hands, but I decided not to linger, thinking I had been blessed enough for one day. Something told me that if I kept doing a good job, I would meet the president soon enough.

I called Heather from the office of the president's secretary, Betty Currie, asking her to come grab me, still not believing that I could be so lucky as to have been asked out by this blue-eyed blond who looked like a Kennedy boy himself.

BEAUTY. Like the famous optical illusion of the old, haggish woman and the pretty young girl inside, I never knew what a man saw when he looked at me. I had grown up in Detroit and its suburbs. If I traveled south of Eight Mile, boys hollered for me—for my light skin and soft hair. If I returned home to Troy, it was hit or miss; some older

---

* This name has been changed.

boys noticed me, while others spat insults and jeers. My dates were usually with college guys who had a stronger sense of self, who knew what they wanted in a girl and didn't feel affected by peer pressure. Though I wasn't ignored, it was clear that I was not an orthodox beauty, the form of girl boys were supposed to like. No one ever made me his girlfriend.

As a child, I'd had it worse.

"Are you a boy or a girl?" one classmate asked when I was a fourth-grader going to school in Madison Heights, Michigan. Recess time, and my friends and I would run for the playground. All I wanted was to climb the jungle gym and feel my weight hang from my hands and kick my feet.

"So what are you?" he asked.

The freckled boy stood in front of the jungle gym. I felt stung and unsure what to do. The boy stood taller than me. I stared into pupils as black and hard as hockey pucks. I never fought in school. I just wanted to play. Now I had to turn around and leave before he said more.

"Mophead!" he called after me, to the back of my curly head.

My skin, brown like packing paper, was different from his. The usual story. But compounding the problem was the matter of my hair. Until age seven, I'd had fine, soft hair; then suddenly, as I turned eight, it became a thick, frizzy curl. My mother, a former Kansas farm girl who had no experience fixing hair with any African in it and who worked midnight shifts forty hours a week, decided that the answer was to cut it. Short. Really short. My curly hair looked like the kind of tight permanent that old ladies used to get. We kept it that way until fifth grade, when I finally started to blow-dry it straight and let it grow.

Until then, I resided in a witness protection program of sorts for females, an androgyny foisted on me early.

Fifth grade was when I started to like boys in my class, but they had little use for me. Middle school felt like one unrequited crush after another. But in eighth grade, a miracle happened. A beautiful boy from another school chose me over my blonde suburban peers. I felt as if a natural law of the universe had been broken. This happened in Washington, D.C., on a school trip. My friends and I spotted a blond boy from Sterling Heights, also there on a school trip. A boy I would later learn had a poster on his wall of Apollonia, Prince's love interest in his autobiographical film *Purple Rain*, and who had a father who raged at him regularly, calling him a "nigger lover." My girlfriends were pretty and popular, but he chose to talk to me, to ask for my number, to ask to see me back in Michigan. I suppose I looked more like the girl in his poster than the other girls in my clique did.

Five years later I was getting a lot of attention in the office, and I lapped it up like a puppy. When I walked in wearing my blood orange dress with the matching suit jacket, three strands of pearls around my neck, Heather stopped what she was doing to look at me. "It's going to sound like a 'Stacy love fest' soon," she said, shaking her head. I was embarrassed but secretly pleased. A month in, and I already felt like Heather's pet, and such pronouncements felt like being scratched behind the ears. But NBC's *Today Show* anchor Katie Couric would compliment me on the color, too. It was one thing to be noticed by Heather. But to be noticed by network stars? Divine.

Then Charles called. Lunch turned into dinner. A good-night kiss accelerated into yes, you can come upstairs. Another night out ended in his bed, in his shared condo in Alexandria. Again. Then again. This boy, this man, was so calm and collected on the outside. But inside, I sensed a powerful core. A life force. I was deeply attracted to him, and I felt as if we were tumbling down a black hole, but in slow motion, letting us fall into something that I hoped was love, or at least its tender beginnings.

Maybe I had finally found my boyfriend. Found him in a garden that no longer had rows of rosebushes but still had the pretty name.

* * *

IF GEORGE FLATTENED HIS FEELINGS to work in the White House, I flattened my voice when I answered his phone. Cool, contained, that's what the job called for, so I learned to make myself small to fit into the intern nook. "Hello, this is Stacy," I said over and over, answering the phone in my low voice so as not to disturb Heather. ABC political director Mark Halperin, who I thought sounded warm and friendly over the phone, complimented me on my voice. That made me happy, even though he complimented the affected me, not the real me, the one who could barely contain her enthusiasm.

Heather and I spoke about voice often. "Stacy," she said, "you need to improve your enunciation. You speak so quickly sometimes that you are not always understood." I listened. In high school, it had been a running joke that my best friend, Peggy, was my interpreter because I spoke so fast and only she could catch all that I said. But what had been cute in high school was not cute in the White House. Heather suggested I take elocution lessons. She suggested this seriously: I should find someone to help me pronounce words better.

I flashed back to the speech therapist in elementary school and being taken out of class to address my stubborn *l*s and *r*s. To be singled out like that made me feel self-conscious, but I followed the woman to her office and did her exercises. I suppose they helped. And I suppose I could have been insulted by Heather's suggestion, too, but I wasn't. She offered me advice that many were not caring or observant enough to offer. Once I returned to school, I signed up to work with graduate speech therapists at GWU, in their lab. I learned how to erase glottal fry and how to stop slurring through syllables. Maybe Heather thought of me as her Pygmalion project, but I felt lucky to have her attention.

* * *

EVERY DAY AT WORK I glanced at George's schedule. Heather printed out two lapel-sized versions: one he kept in his breast pocket: and the other she kept on her desk. She spent the whole day assembling, refining, and rearranging his itinerary, which was based on the president's and other participants' schedules and whatever erupted that day. The work looked like refueling planes in midair, and I was glad it was hers and not mine. It made me think I was a long way from being able to master her job. I could barely keep on top of all of George's meetings.

Meetings. Harold Ickes later said that meetings were the stuff of White House life, that the trick was figuring which to attend and which to skip. Countless daily meetings were devoted to some aspect of message development and implementation. Long-term message, short-term message, message of the day. This administration, like the ones that had come before, spent a lot of time crafting the rhetoric and imagery that would sell its agenda to the American public. Stories mattered.

Pictures mattered more. The White House knew that people remembered pictures long after they forgot a correspondent's voice-over, no matter how damning the narrative. Ronald Reagan was the master. When unemployment spiked in 1982, he drank beer in a working-class pub and made sure the press pool photographed the images. Message: *I care*. When Reagan's budgets slashed education spending, he toured public schools, making sure he was photographed next to happy schoolkids for the evening news. Again, the message those nightly pictures conveyed was compassion. It didn't matter if the correspondent Lesley Stahl attacked his hypocrisy during her voice-over. What the voters remembered were the pictures. "Optics" was key.

So was "verbiage," the pithy phrasing used to articulate a message. "Putting people first" or "Mend it, don't end it" or "Building a bridge to the twenty-first century." Words on banners above the president's head or, if printed on the backdrop behind him, suspended like snowflakes over his shoulders. Journalists acted as filters, interpreting whatever it was we wanted to say. But if our pictures made the papers, readers soaked in our pure message without interference.

I never sat in on the message meetings. Neither did Heather. She scheduled them. She kept the machinery moving.

"Where's George?"

Rahm! Like a whipcord behind me. He popped into George's open office, popped back out: "Where's George?" he asked again. Heather covered the mouthpiece of the receiver in her hand and whispered, "Roosevelt Room." Rahm nodded, knowing already which meeting that meant. He didn't move. He stood there, his taut body just two feet away. A full moment's pause, when he smiled at me before he flew out the door.

Rahm barely knew me. I'm sure that to him, I was just the intern who answered George's phone. But did that thirty-four-year-old man know what his smile could set off in the mind of an eighteen-year-old girl?

A crush was born.

"You're so diva-ed out!" the pretty stranger said to me as I pushed in next to him at the crowded bar. "Thank you," I said. I loved the attention of this David Bowie double, even if he had no interest in me as a lover. We were at Tracks, my favorite nightclub, on one of its gay nights. Behind me were Justin and Mike, waiting for me to bring back our drinks.

I felt so happy and free among my beautiful new friends. Justin was tall, with suave good looks and a touch of Native American in his face. Mike was shorter, Italian, with an easy smile, the kind of guy you always wanted to hug. Justin and Mike were boyfriend and boyfriend. In my high school, *no one* had come out, and I had never had openly gay friends. Washington was a revelation. Justin and I had met in my dorm, where he worked as a community service aide, checking in guests. Having just recently transferred from the University of Oregon, he was making new friends, and we hit it off, going out and talking on the phone. Justin would laugh at me when he called me in the White House and I sounded so official. If he called me from his job at the World Bank, he sounded official too, and we laughed at each other, two masked people playing our roles.

By summertime, Justin had come out to me, and soon we were bug-eying the drag queens at Trumpets and singing along with the televised video sirens at JR's. In Washington's gay bars, no one corrected my diction or told me I didn't look pretty enough or white enough, and the mood was fun. No wonder you found so many straight girls in these bars, I thought. But we also went to straight dance clubs such as Fifth Column or The Vault and in our drunken silliness would later run up Pennsylvania Avenue, laughing our lungs out as we tried to make the White House's sound-activated lights brighten the front lawn. If that didn't work, Justin would slap his hand over an infared beam embedded in the column and we would run away, trying not to get reprimanded by any uniformed Secret Service on patrol—before I had a blue badge and actually knew some of them.

I stood there with the boys' drink money in my hand as a CeCe Peniston mix played loudly all around. "With brown cocoa skin and curly black hair, it's just the way he looks at me, that gentle loving stare . . ."

Like Rahm Emanuel. I wished he were there. To be that close to Rahm, in a place where touch was so easy, with every corner dark, beckoning, and available. Delicious just to imagine. I saw Rahm every day in the West Wing, and now I was in seventh grade all over again, yearning over his senior high school yearbook picture:

the beautiful, to-die-for Rahm. Lately, whenever he saw me he said hello, leaving me stunned that this gorgeous man was paying attention to me. Rahm talked to me as if I were a true prize, the kind any red-blooded man would fight for.

And Rahm knew how to fight. Fellow Clinton campaign veteran Paul Begala called him "Rahm-bo," this man rumored to have both served in the Israeli army and danced ballet. A dancer and a fighter, I thought. That was my dream doll. The yin and the yang, flowing, the midline taut with sweet tension. Testosterone. Passion.

"Pretty thing," said a nattily dressed black man who kept on walking. "You always get compliments," Michael said. I made a face, dismissing his words, for it certainly wasn't true. But I suppose that night I did stand out, with my boots and short black dress, my wavy bob and my red-lipsticked lips. But what did it matter? These guys were gay. I wasn't thinking that to most of us, any attention is good attention.

Step here, and you're attractive. Step there, and the searching gaze slips right past you. Whatever physical beauty I had, I felt so outside of it, caretaker to some force that in the end felt useless. At eighteen, I remained the only one in my circle of friends—including Justin—who had never had commitment with a boyfriend. Charles took me out, but I knew that he didn't feel deeply for me, that he might never love me. What good was this power if I couldn't get that?

We maneuvered through the crowd until we made our way toward the steamy back of the main dance floor. I smelled dry ice and Joop! as men took off their white T-shirts and danced. We danced too, together, as if we were each other's dates, as if I had two men to myself. I loved it. I hated most straight bars because I couldn't bear the macho guys who came on to girls with the finesse of cannonballs. Tracks was a meat market, too, but for the boys. I was a voyeur there, surrounded by men who turned me on without touch. I soaked in their energy without risk of being seduced and abandoned or infected. This was paradise, I thought. The euphoria of entrée.

THE NEXT MONDAY, Heather asked me about my weekend, and whether I had seen Charles. I said we had talked but I had spent the evenings with my friends. "Your birthday is coming up, right? Will you do anything special for it?" I didn't know, I told her. I had mentioned my birthday to Charles and I hoped he would make plans. Heather smiled and said that for sure, a girl like me would be remembered. I thought back to how she had led me to the Rose Garden, where I had met that boy, how our love seemed to be fated.

Then I thought of our last phone call, the one that had lasted five minutes, when I had struggled to connect with him and we hadn't had much to say. If this were the real deal, wouldn't it be as it had been with Peggy, my best friend in high school, back when we had talked and talked for hours and sometimes kept the phone at our ears as we fell asleep? *Don't go yet. No, don't hang up.* When I had left for college, Peggy and I had typed up notes of our favorite memories together and given them to each other, notes I have to this day. If I cried, she knew how to hug me. I thought love with a man would be like that, too.

Don't worry, I told myself. This birthday, the first one celebrated far from home, should be amazing. I stoked that anticipation as I pulled out another letter and began working, eager to fill my day's quota.

4

Nineteen Candles

I WAS LIVING BY MYSELF FOR THE FIRST TIME IN MY LIFE, IN A SIX-hundred-square-foot efficiency my mother had rented for me after she swooped in like Superwoman and helped her finals-fried daughter find a new place, pack up her stuff in Thurston Hall, and move out. We had looked at one apartment on Pennsylvania Avenue and Twenty-fifth Street; my mother had taken one step inside, smelled the combination of mothballs and roach spray, and declared, "No, you're going to have problems."

Instead, we chose Potomac Plaza, across from the Watergate complex. My mom liked that the front desk was staffed twenty-four hours a day. Most of my neighbors were older: married couples, retirees. Primarily white, but I had seen Indian women and a few Asians. Sedate professional people, making for a quiet building. Fine with me, I thought, because I was never one to have many friends over. I liked my private space to be private, a place that, even if small, was inviolable.

My mother paid the rent. Expensive—$650 a month on her nurse's salary, in addition to what she contributed to the household and for my baby sister. I don't know how she did it. My parents' finances remained mysterious to me, but it was clear that my mother paid for many things, despite making less than my anesthesiologist stepfather. I think my mother paid my rent for her own peace of mind and could do so given my full academic scholarship. She had eyeballed the apartment, thought it was safe, and thought I would be, too.

My mother drove me to the IKEA in Virginia to buy furniture. She never said no when it came to housewares or most anything I asked her, so I tried not to ask her for too much. I had a double bed, a big black desk, a TV, a carpet, lamps, a few chairs, dishes, a microwave, a small dining table. A few Billie Holiday CDs and wineglasses—enough to signify the kind of adult I wanted to be. My possessions would never compare to those of the pampered daughters of Columbia Plaza, the New York/New Jersey girls whose families subsidized more of their lives than mine did, but I never felt that I wanted for anything.

I hated asking for money. Each time I picked up the phone to do so, I ached for the time when I wouldn't be so financially dependent on my mother, for the time when a call from me would not set off the bell in her head: Oh, how much is this call going to cost me? I hated the idea of being a princess, of being spoiled in any way. But clearly I was one, at least to my mother. Not the kind to dress up, to show off, as if only a doll; the kind to nurture and protect and help succeed.

My birthday approached, and I wished I could have gone home. During my first year in college, I had flown home for my high school homecoming weekend, Thanksgiving, Christmas, and Easter. For someone who claimed to be so independent and happy in her new big city, that was a lot of flying home. Work needed me, and I would stay put in D.C. Plus I looked forward to spending my nineteenth birthday in the West Wing, a gift unto itself.

JUNE 18 ARRIVED. I woke up with a happy start, as if my parents had just packed the car for a day trip to Cedar Point amusement park and it was time to go. I lay there alone, soaking in my anticipation. My double bed pressed against the window-lined wall that let me look straight into the branches of a tall, leafy tree. The sun rose on the other side of the building, so the light had a watery quality, clear gray like the river nearby. I loved how the windows framed one square of nature at a time, allowing you to concentrate on what could be overwhelming when you're out in it and there's so much stimuli and you need to go wherever you need to go.

The previous week, I had seen my name and birth date written on a Post-it in Heather's Rolodex. I knew that Heather had asked George to call one of the volunteers on her birthday, a woman who worked primarily in the OEOB. George and the volunteer had met only once before. I'd worked in his personal office almost every day for a month, answering his phones, typing his letters of regret. Would Heather arrange a moment for me? I tried not to get my hopes up. But oh, I hoped for *some* sort of acknowledgment. A nice "Happy birthday" from George and Heather would be heaven.

So would a kiss from Charles. Or a note. Some gesture to let me know he cared.

Showered, hair done, bag over shoulder, I scoured the apartment for anything left behind, telling myself that I needed to face reality: three weeks after the Rose Garden meeting, Charles was not my boyfriend. Not even close.

But we slept together. He ran marathons, and I loved touching the muscles of his thighs and his calves. On bare feet, we stood eye to eye, his body hard and compact as he held me in his double bed with his door locked, his roommates only walls away.

We spoke regularly, but our talks still felt like laps on an ice rink. Once he shared part of a poem he had written on the Metro, lines about casual sex and a hard shell uncracked. I couldn't tell if it was a rebuke, a plea, or a statement of fact. Yes, I was excited when he called, just as I was when we kissed, but I wanted explorations to where fingers couldn't fit.

FEW PICTURES of my early childhood remain. I treasure each rare image, and one of them is of me, sitting alone at a play area table at St. Philip's preschool in Detroit, in front of a round Snoopy cake large enough to feed my whole class. Why I am alone, I don't know. I have made a birthday wish, for sure. Maybe it was for more Legos. Maybe it was for my parents to stay together.

I lived in Detroit proper those first years. As a child I felt sealed in, surrounded by family and factories and houses like faces, including the ones left standing with their burnt-out eyes and boarded-up mouths. The city hollowed itself out, one see-through structure at a time. The unemployed stood in long lines, but the adults felt pride in Coleman A. Young, the new black mayor elected just seven years after National Guard tanks rolled down Rosa Parks Avenue before it was Rosa Parks.

None of this mattered to me as a little girl. I loved being in the city! Look at the old pictures of me, so bright-faced and curious. Open like the summer windows as neighbor kids ran around me and I ran around them, crying, laughing, our brown skins shining. I was happy to meet people and happy to be met.

I found stillness alone in my room, no matter the constant moves from apartment to apartment, be they in Indian Village or Cass Corridor or downtown in the Madison Hotel or on the Westside, in my paternal grandmother's duplex. I sat on Nana's stoop, and she drilled me on elocution, trying to unstick my *la* from my *ra*, then sending me off to the corner store for a BombPop. "Watch the cars," she warned, and I would. I watched the long cars heavy with metal and V-8 engines. From their radios I listened to songs by Stevie Wonder, Marvin Gaye, and my father's favorite: Blood Sweat and Tears. Whenever I heard "And When I Die" I was with my father again, and I was happy.

My father's full name was Edward Stanley Parker III, but friends and family called him "Butch." I didn't see him much once my parents separated. He was too sick. A full-blown alcoholic, he was burning through his gastrointestinal tract and was in and out of the hospital. During the last months of his life he lived below my grandmother in her duplex. He listened to music when I visited, listened much as I would, growing up: incessantly, as if fed by the songs. He owned a reel-to-reel and an eight-track machine. Butch held me in his lap and sang "And When I Die." *There'll be one child born in this world to carry on.* Those special nights when he pulled me up onto his La-Z-Boy, I felt serenaded, with the moon visible out the high window glass and me in my daddy's arms receiving his full attention, having no idea what death meant, that this was how he said good-bye. In the morning, my mother would fetch me and return us to our apartment at the Parkcrest. I resisted. I loved my mother but I wanted my father, though his house was no place for a child.

Neither was the Parkcrest. The four-story, scalloped apartment building lay between Lafayette and Jefferson avenues, the crumbling thoroughfare that runs parallel to the Detroit River. If you keep driving east until you reach the tony, monied, dripping-with-self-regard Grosse Pointes, Jefferson Avenue kisses up close to the water. But we never drove there. In our part of Detroit, side streets and warehouses separated us from the bright river, its blue the result of chemical dumping and whatever else people did when nobody stopped them.

My mother knew Parkcrest was no place to live, but the apartment was all she could afford near her nurse's aide job at Deaconess Hospital. Mice and roaches infested our one-bedroom, yet the landlord had the nerve to tell her that it was children who were unwanted. My mother went to the local ACLU. She found a lawyer willing to confront the landlord with a rental discrimination claim. The landlord backed down.

From then on, the landlord called me the Princess of Parkcrest.

If I was a princess, my mother was an angel in nursing whites, her black hair pulled back, her manner equal parts heart, pragmatism, and survival instinct. My mother refused to give up or frighten just because she was a white girl alone in the city. My mother's Detroit was a town in rust-belt decline, the downtown streets cracked by the cold of the hard icy months followed by the summertime heat. She walked to the bus stop alone, beneath advertisements spanning tall building sides, the paint chipping off the bricks.

My city and my father grew sicker, but my mother persevered. Having served as a nurse in the U.S. Army, she used her Veterans Affairs credits to attend community college. She kept working as a nurse's aide in the emergency room, doing the tasks that so many other hospital staffers would rather avoid: cleaning up after others' sick, broken bodies. She enrolled me at St. Philip's preschool. Nana helped as best she could, but my mother kept it all together, even when I begged for her attention as she sat cross-legged on the couch, a heavy textbook in her lap. I would pull out the brown ribbon of her Blondie and Bonnie Tyler tapes. If that didn't get her attention, I would sit on her textbook, then fall on it, belly first, my brown eyes looking straight up into hers. I would stay there, sprawled across the pages until she lifted up my reluctant body by its full weight.

Funny that my first memories of politics are from Henry Ford Hospital on West Grand Boulevard, practically across the street from the old Motown studio. Take me to the hospital, I'd say. I wanted to go. I wanted to go see my daddy. Take me to Henry

Ford with its hallways like airport terminals, with its busy doctors, nurses, and visitors. Busy, busy, so much motion—but not so much that people didn't stop to look at me and smile or touch my hair. My grandmother worked at Henry Ford as a neonatal nurse, and I received a lot of attention.

My favorite place was the cafeteria. Nana purchased balanced meals, but I only had eyes for the chocolate pudding. When we wandered between the tables, I already sensed that there was a hierarchy, with the doctors sitting there, the nurses there, and the orderlies over there. So much to learn and negotiate.

"Come on, baby," my mother would say when she met us at Henry Ford after her Deaconess shift. Together we would move through the tangled, blinking, beeping maze of legs and purses and wheeled beds and onto the crowded elevator. She would hold me tight next to her white work pants, her hands on my shoulders, after she smoothed down my curls.

My mother and grandmother would enter the room first. My mother held my hand, but I broke free and ran to the bed. Daddy was awake. He looked different, but I didn't care. It was him, just smaller. And skinnier. Still long, but the pajamas fell from him now.

My mother helped me up onto my father's bed. He smelled like himself, plus a thin medicine scent. I lay there next to him as a young wife might. I lay there next to him until I was pulled down by somebody else's hands—whose, I don't know.

By then my father had become too sick to work. He had once aspired to practice law, maybe even politics, but in 1968 he had dropped out of Wayne State University and enlisted in the U.S. Army. He'd walked right up to the enlistment office and volunteered. He was rebelling somehow. He was tearing up his dreams, his family's dreams, in everyone's face. After serving two tours in army intelligence, he returned to Detroit physically intact, but when he tried driving buses for a living, he drove them into curbs. When he drove cabs, he clipped them against other cars. Those jobs, so

beneath his pride and abilities, were lost. He never confided anything in his small daughter that I can remember, but I imagine his
actions as being those of someone deeply in pain, with an ache and
an anger that he tried desperately to self-medicate. I was born and
we moved from apartment to apartment, from bad to worse, until
Butch and my mother separated, until he finally died. I was six.

The death certificate states the cause as internal bleeding. But
in truth, he had drunk himself to death by age thirty-one. It was
1980. He left behind a widow, a daughter, and a mother forced to
bury her only son.

Life changed dramatically for me when my father died. My
mother met Rufino, an anesthesiologist from the Philippines, and
our lives grew more stable. Suburban. He took us steadily up I-75,
each move from apartment to town house to new house a move
up. In 1982 they married in a civil ceremony while I was at school.
However, their marriage resulted in no felt difference in our home.
It was still the three of us, together.

Except for when it was just me. My parents worked long
hours, and I was on my own a lot. I loved drawing horses and
making mix tapes from the radio. I placed my first radio request
at age six ("Sleeping Single in a Double Bed," by Barbara Man-
drell). FM radio was my surrogate parent cooing love songs like
lullabies, making me as happy inside as two tickling hands. With
the radio, mood management was just a dial switch away. However, the happiness of solitude seems very chicken and egg: Did I
prefer being alone? Or did I simply learn to take pleasure in what
I could not control? I don't know. Regardless, I was happy. I had
a second father. A man who provided for us. He created the safe
space that allowed me to grow up unbothered and unmolested. I
could follow my dreams without worry that the floor would fall
through. And the best part: my mother gave birth to my sister,
Elizabeth, my world-dominating brainiac of a sister thirteen years
my junior. We're separated by time and by gene pools, but ask us a

question about politics or pop culture, and you'll find us finishing each other's sentences. Once I told someone that she and I were "half sisters," and she immediately added, "But we round up." I totally agree.

My mother, Rufino, and Nana provided me what Butch could not. Security. Consistency. Love. Together they nurtured a girl confident enough to speak up for herself and others, brave enough to enter the strange new situations where I found myself in Washington, D.C.

I DIDN'T YET HAVE A HARD PASS, so I walked through the Seventeenth and G Street entrance of the Old Executive Office Building. I cut through the cavernous hallways of the OEOB, across the asphalt of West Executive Drive, up the stairs to the West Wing. The marine was there. That meant the president was, too. The marine opened the door for me, and I thanked him as I entered the West Wing lobby, where Debbie Schiff, the West Wing receptionist, would call Heather to come pick me up.

The lobby I waited in would not floor you with grandiosity. The ceilings were high, but not like those of a palace or even a bank. It was simply an elegant room, furnished conservatively with the best of what is reasonable to spend of taxpayers' money. Dark wooden tables surrounded by blue chairs and yellow couches. Buttermilk walls, white moldings, and dark wooden doors. Large paintings of historical scenes in big gilded frames. Though pleasant, this room felt less about decor and more about the anticipation and tension of visitors who waited to see staffers or the president himself.

Debbie picked up the phone to call Heather and let her know I was there. A pretty fortysomething Dallas blonde, the West Wing receptionist always looked put together without looking fussy. I loved that she kept a glass container of M&M's on her desk that passersby, such as me, could enjoy. In her drawer she kept the coveted boxes of presidential M&M's with the POTUS signature, one of the coolest gifts to give friends and family no matter their politics. She'd given me a few boxes already. Debbie also controlled the extra tickets for the president's box at the Kennedy Center. Heather mentioned perhaps going to see George's girlfriend, Jennifer Grey, perform. Maybe Debbie knew it was my birthday?

Relax, I told myself. Surely Heather was planning something cool. Heather could go quiet when she was overwhelmed or irritated by an ill-timed question on my part, and the mood could take time to pass. But I never doubted her gratitude. She complimented my letters often. She even complimented my person. "Look at her," she gushed one day to James Carville, who, despite not being White House staff, called George often and sometimes came to visit. "Look at those almond eyes that go up." I blushed as James stood there agreeing with her. As if on cue, my crush, Rahm, entered the office. "See those eyes!" Heather cried, and Rahm agreed with her, too, making me burn with a grateful happiness indistinguishable from embarrassment.

Heather said nothing as we entered the office. I looked on my chair, in my in-box. A surprise birthday card? No. Just a thick folder of George's personal correspondence that needed responses.

What had I expected, a balloon drop? I put my stuff down and booted up the intern computer. The morning light fell in from George's open office door, into our bright space made brighter by the white-painted wood cabinets and credenza that reminded me of the Sears children's bedroom set I used to have, the one Rufino had bought for me when we moved into our first suburban apartment.

I'd thought she would remember. Heather felt like such a force in my life, and for her to treat this like any old day was disappointing. The phone rang and I quickly settled in, taking the calls, hiding my hurt behind my work voice. At this point, I knew how to handle 90 percent of the callers. I knew when to interrupt George, when to take a message, and when to ask Heather. Learning this made my job less harrowing. But not less busy. Letters needed to be finished. No time to feel disappointed.

Ten a.m. passed. Eleven a.m. passed. I looked at the clock at 11:30 a.m. on June 18, and nobody—including my mother—had acknowledged that it was my birthday. Was this the year my life was going to morph into a John Hughes movie? I felt more like a *Pretty in Pink* Molly misfit than a *Sixteen Candles* girl, but each time that I called home to check the machine, I heard the same taunt: zero beeps meant zero messages. My first birthday spent away from home, too. I guess when you leave, you really leave, I thought.

I worked through my file of assigned correspondence. A Greek Orthodox church in Florida wanted George to appear at its Greek Festival. It made winning references to his family members, attempting to show connections—

I spied George's thick stack of clips on Heather's desk opened to a *Wall Street Journal* piece. I saw the sketch portrait. Even upside down I could tell it was Rahm.

"May I look at this?" I asked Heather. She nodded, blowing some air in a way that said the article might be trouble.

It was a full profile on Rahm. I saw the word "cutthroat" and grimaced. I guessed that could have been used to describe Rahm, but I never saw him that way. A body for movement, yes—everything about him push, push, pushing, moving faster, moving forward. I'd just learned the Yiddish word "macher," meaning a man who makes things happen, and that seemed to be him. I sensed the striver in him but not a man who sought only money or power and lived to snatch for it.

But I was just another man's intern. I never sat across from him at the negotiating table.

I felt a presence rush up behind me and there he was. Rahm's eyes glanced down at the article, glanced back at me. He raised his eyebrows, then popped into George's open office. Popped back out and asked, "Where's George?" Heather covered the receiver and whispered, "Roosevelt Room."

But he stayed put. We started talking. All that energy coming right at me. Subconsciously, I must have been fishing, but at the time I would have sworn up and down that I asked for Rahm's birth date just to see if we were astrologically compatible. So I asked him.

"November 29," he said. "When's yours?"

"June 18," I said.

"That's today!" said Rahm.

"Well, yes, it is."

From the corner of my eye I caught Heather looking down at her desk. She acted naturally, but I saw her glance at her Rolodex. It's okay, Heather, I wanted to say. I sensed she had wanted to do something for me. She'd just forgotten.

"Have a wonderful day, okay?" Rahm said.

Rahm may not have been the character Jake Ryan, kissing me over the birthday cake à la *Sixteen Candles*—one of the greatest teenage wish-fulfillment scenes ever captured on film—but I felt pure warmth inside. The feeling of being singled out, as if adored.

Yet I wasn't delusional: Rahm loved another woman, his fiancée. The boy I really wanted was Charles, and he was MIA.

AT LUNCH, I walked to the ATM in the OEOB, outside the ground-floor credit union. In this building, I could walk freely. I saw David Leavy finishing his transaction. I liked David. A junior staffer in communications with a cute Judge Reinhold smile, he had always been nice to me. "Hey," he said, and we started chatting. I told him it was my birthday. He hugged me, asked what I was doing to celebrate. I mentioned Charles, and he immediately asked, "Has he sent you roses?"

"No," I said.

David's eyes cast their judgment. "Well, there's still time. But if no roses tonight? I don't know about this guy."

I wrapped up the conversation quickly. Despite the beauty of such a gesture, given that we had actually met in the Rose Garden, what was the chance that Charles would actually spend money and forethought on flowers? No guy had ever sent me flowers. At that age, I thought they were a cliché and a waste—sixty dollars that could go a long way on something lasting. I would rather have a card inscribed with a heartfelt note. But Charles did not love me like that yet. This I knew.

The fact was, neither Rahm nor David could staunch my disappointment. One p.m., and no one had remembered my birthday.

THE DAY CONTINUED like any other West Wing day: glimpses of the president, agents, and staff moving up and down the hall; stressed-out callers; and journalists hungry for George's time. I typed my letters. I answered the phones. I briefed Heather on how the correspondence operation was going in 415 OEOB. When it was pushing 4:00 p.m., my disappointment shifted into sadness. Forget Charles. Forget these White House folks, I thought. How could I have let myself connect my happiness to their attentions?

"Stacy," said Heather, "let's take a walk."

I looked at her and immediately nodded. If she told me we needed to go out back and build a plane out of Popsicle sticks, I'd have said yes before ever dreaming of betraying that her request might have been beyond the scope of my skills.

I followed her out the door, aware that the phones would be un-manned and whatever we were doing would have to be done fast.

We walked three yards and stopped. Heather stood directly in front of the Roosevelt Room. "You're going to watch the radio address," she said, still firm in her voice but with light in her eyes. "Happy birthday," she said, hugging me. She wore a suede vest that was soft to the touch.

Heather led the way as I joined an invited audience of fifteen guests that would watch the president record his remarks. The Roosevelt Room provided an even more intimate setting than the Rose Garden: guests sat up close and personal, just feet away from the president, guaranteeing moments of eye contact with him, if not a handshake and hello. "See you when you're done," Heather said as she pivoted back to the office.

I took a seat in the back-row chairs set up in this storied room with the ornate western feel befitting its namesake, Teddy Roosevelt. George attended meetings here all the time, including his senior staff meeting each morning. I saw Dave Anderson, who coordinated the radio address logistics. He smiled and kept working.

The seats around me filled up. Parents and children—who all seemed to be friends and family of the president or of his top staff—seemed awed by the occasion, including the dressed-up young children, who couldn't have had any real idea as to how special this event truly was.

The president's communications strategist, David Dreyer, walked in. One of the first staffers I met when I worked in News Analysis, he had a dark beard and warm eyes. He walked over and wished me happy birthday, giving me a huge hug, then departing again. I beamed inside. Heather must have told him.

We all sat there, waiting with that tingly anticipation you get when you know that POTUS is about to walk into the room. It felt strange but delicious to be there as a member of an audience and not someone running to try to make sure everything worked right: to be served, as opposed to helping the chef. I saw Heather come back in, grab Dave Anderson by the sleeve, and whisper something in his ear. She left without looking at me, gone as quickly as she had arrived.

"I'd like to announce something," said Dave Anderson to the assembled group. "Stacy, would you stand up?"

What?

"Today is Stacy's birthday, everyone. She's an intern here. Happy birthday!"

Everyone applauded. This wasn't John Hughes anymore; this was the fruition of little-girl dreams. The assembled guests turned around in their seats to see the girl singled out. Terribly embarrassed and happy, I sat back down.

George walked in the side door and strode toward me. "Happy birthday, Stacy," he said, extending his hand. People turned around once more—now *really* happy because they were seeing a bona fide star. "Thanks for all of your help," he added, smiling, earnest, before he turned and left the room.

Heather appeared in my mind's eye, orchestrating all from her black chair, the director making actors appear with the authority of her commands.

Then the back door opened, and in walked the president. Everyone stood up, clapping, resisting the urge to lunge toward him, to claim his undivided attention. Look at him, so tall and handsome. I didn't have a crush on him, but I could see how others did. The president greeted nearby audience members and took his seat in front of the microphone. In walked the official photographer. Other staffers, including George, lined the wall.

The president delivered his address. I sat there, thrilled that this was how I was spending my birthday, just feet away, like a VIP.

Once he finished, Heather slipped in the door. "Stacy," she whispered to me. I went to her. She walked me near the president, who was talking with a family from Arkansas. The president turned to us. "Us" had turned into "me," for Heather slunk away.

My God, his eyes. That powerful blue looking right at me. He extended his hand. I had no time to think. "I'm Stacy Parker," I said. "I do George's mail." I felt the heavy *click, click* of the camera near us, locking the moment on film. "I know," he said, his grip firm. Did he really? How did he know? He thanked me for my help, and though the moment was short, it seemed to elongate, for he seemed to linger in the moment as much as I did.

Then it was over. He talked to another couple, who were smiling, waiting their turn.

Oh, my God! I wanted to scream.

Maybe this was heaven for the president, people wanting to see him and understanding that they would get a moment of his time; then he would move on to the next group and drink them in, too. All I knew was that it was heaven for me.

I rushed back to the office, barely able to contain myself. "Happy birthday, you," Heather said as we hugged again. She felt small but powerful in my arms. "This was the best birthday I've ever had," I said. "I'll make sure you get that picture," she added.

She did. You should have seen the look on my face. Surprised. Delighted. Unbelieving that this was how I got to spend my first birthday away from home.

As for the rest of my world: my parents and grandmother called in the evening to send me good wishes and to let me know my gifts were in the mail. Justin called, asking me if I wanted to go to the bar and celebrate. I told him to hold on, that I thought Charles and I might still go out. Peggy called and so did Christina, my good girlfriends from home. I felt remembered by the people who really cared about me.

I waited to hear from Charles. He did not call. My Rose Garden dreams felt patently ridiculous. In my journal, I vowed to myself to keep vigilant, to keep discerning fact from fantasy, not to be fooled by fantasy, to know that it was so easy to imagine more of a relationship than what actually existed.

At 9:00 p.m. I decided to call him myself. I told him matter-of-factly that this was my birthday. He apologized, acted doleful, and said he hoped I'd had a good day. I said I had. I felt disappointed in a deep place, the kind of feeling that no amount of self-chastisement or rationalization would erase. I hung up the phone, mad at myself for having waited around for him. I cried for a few minutes, sitting on the floor, my back against the bed. I was laden with the realization that I was always going to want men to be there for me, no matter how well everything else was going in my life.

But then, for a moment, I forgot about Charles, for I was looking at my hand that only hours earlier had shaken the president's hand. I thought of the Roosevelt Room.

Do you know what it is like to suddenly be the celebrated girl? To be spoiled with attention by people who are celebrated themselves? To have those people respond to a desire so deep inside you, so primal, that you yourself can articulate it only in the pages of a journal as it bubbles up in moments of epiphany?

Dizzy. The kind of feeling you don't tell people about, because they might resent you, because not all of us get these moments right when we want them. Later, when people asked me about my White House life, I showed them the Roosevelt Room picture. But if pressed, I kept my feelings of Hollywood wish fulfillment to myself.

5

White Gloves

I rushed down G Street because I was thirty minutes late already and oh, God, each minute Heather waited distended like taffy. I'd studied German in school and learned that the Germans never said "late," they said "zu spät"—too late.

I walked into the West Wing office and looked first to Heather, then to the clock. Forty-five minutes past. Heather sat with the phone pressed against her ear, her head tilted down, her black bangs long, a fence.

She refused to look up.

Another line blinked red. I whispered, "Do you want me to—" but she didn't answer. I slipped into the chair, booted up the computer, stuck my bag beneath the desk. I assumed she didn't want me to handle the blinking call. I sat there, feeling awful and afraid of what might happen next.

Julie had warned me that Heather was moody, but I felt I deserved this. Heather treated me like staff, and I couldn't fall back on the excuse that I was "just an intern" when I screwed up. At least I would not want to: that would widen the distance between us.

For Heather's needs were now my own, and I knew she depended on this time, these precious morning hours, to get her most important work done. How could she focus if she was stuck answering call after call because her assistant was late? She drafted and managed George's schedule—a constantly shifting, reconstructed, delicate thing, requiring constant judgments and negotiations with other assistants. I knew how hard it was just to draft a letter when you were always picking up the phone, your concentration constantly ruptured, recomposed.

Worse, some calls were more painful than others, constituent calls especially. Mike from Cleveland was pissed that President Clinton supported gays in the military. He wanted to make sure you heard every word of his rant. You tried to slip out, but persistent or belligerent callers would not allow you a dignified exit. Regardless of their tone, we knew we needed to handle constituent calls with care and that George wanted to know if we received several calls on a particular issue. These calls could go on for five, ten minutes. Other lines blinked, other staffers rushed in and rushed out—the million-miles-per-hour world raced past, unstopping, but you were in this twilight zone with a stranger, having to listen, to be pleasantly passive, and if you didn't know how to shimmy out of the calls you became the problem for being too weak. Voters deserved to be heard, but sometimes callers lost sight of the fact that they were speaking with a real live human being, usually low on the totem pole, with no power or choice but to listen. They spoke as if they believed that if they pummeled you hard enough with their words, their policy dreams would come true.

"Good morning," I whispered, trying again. Heather looked up, but her expression remained blank. She sighed silently, taking in or blocking out whatever the caller was saying, tapping the lengths of her unpainted nails against the glass-covered desk.

I pulled the "to do" folder out of the intern slot of trays connected to the wall near my shoulder. The thickest folder of assigned correspondence to date. If I followed my normal pace, they might take me two or three shifts to complete. I vowed to get every single letter finished today. I believed if I worked really hard, she would forgive me.

The phone rang, and I grabbed it immediately. It was the journalist Elizabeth Drew for George. George's door was closed, so I knew not to interrupt.

Heather remained silent. I stole a look at her. No redness to her eyes, so I assumed this wasn't a bad personal call. I'd seen her cry before. She hadn't wanted me to see, but she'd received some news once and gone and sat in George's office with the door closed. Her friend Wendy Smith had come and sat with her. She hadn't told me what was wrong, and I hadn't pried.

"Yes, I understand your frustration but . . . Ma'am, I'm sorry to hear this, but the administration—"

Oh, no. A constituent call! Dread as if I'd swallowed poison and it was spreading through every vein, artery, and capillary. If only I'd been on time, Heather would not be captive like that. It would be me, as it should be. Heather had taught me to take responsibility for all aspects of my job. To never accept inadequacy but instead engineer solutions. Letter responses not moving fast enough? Request a second computer. Understaffed? Request more interns.

Can't wake up in the morning? Use two alarm clocks.

I heard her caller get louder and faster. "Heather, do you want me to take it?" I whispered. Heather shook her head with a short jerk.

She hung up the phone. Her hand remained on the receiver, as if it had taken all her strength to return it slowly and not slam the thing down.

"It's been literally crazy in here," said Heather, enunciating each word and not looking up. "The phones would not stop ringing."

Each syllable hurt me as I sat there, feeling entirely at fault. I vowed to fix this, not unlike a child who thinks she can keep her parents together if she takes on more of the chores. I would just work harder, I thought, and clean out my in-box of special letters, working steadily until I finished them all. No slowing down. No getting distracted. No daydreaming about Rahm. Eyes on the prize, and she'd forget to be upset with me. I'd keep myself as small as possible and not irritate her, even though we sat two feet away from each other in a room like a bright walk-in closet, the walls reflecting their whiteness off one another. Anything to keep her from getting cold with me the way I saw her do with other staffers she felt were unhelpful or inept. I never wanted to be the object of her contempt. Never.

I looked over, hoping to catch her eye. I could see the black lace on her camisole. Not in your face, but suddenly you saw it and thought *oh*. She wore blazers and denim jackets over low-cut, silken tops with skirts and boots, creating a look that let you know she was not just any old Washington assistant or a brains-only wonk but a woman with star quality who always seemed so much taller than she was.

Today she wore an orange linen jacket, the color of autumn, its sleeves rolled up. A long narrow skirt. Boots with heels. And a low-cut blouse, but not so low as to be revealing. Except for that tease of lace when she bent forward. I wanted to tell her she looked nice, but I would wait. The tension hadn't lifted.

"Stacy, before I forget. There's something I've been meaning to ask you. You know how George is giving that intern Q and A next week? We need someone to make the introduction, and we'd like it to be you."

"Really?" I asked. "*I* get to do it?"

"Yes," Heather said. "You deserve it. Write something good."

I sat there stunned. On a day when I had left her in the lurch, Heather looked at me, said, "You will do this," and with words like waves of a wand propelled me into a situation that only moments before had been unimaginable. In a room like a large college seminar classroom with its stadium seating and bright institutional lights, I would speak before the other interns. I would be first among equals, looking out at the rows of young men in ties, young women in suits or dresses, looking like the professionals we wanted so badly to be.

The brown-bag Q and As were a big deal. I myself couldn't wait to see Dee Dee Myers, for even with my access, I didn't see her much. That was the beauty of these events: they brought interns close to the star staffers with whom they normally had little or no contact. They were a chance to get in a handshake, a word, so relationships could begin. Or to ask something clever enough to get a laugh or "That's a good question" from the speaker.

At least, I imagined this was the dream of the true go-getters, the ones stuck in an OEOB communications office reading through several newspapers and magazines a day scouring for Clinton administration hits but never seeing in the flesh the men and women who did the sourcing.

I had to get this introduction right! Pride burst from my heart, made me smile from ear to ear whenever I talked about George and the honor I felt working for him in the White House. This need not be hard, I told myself. I could simply read the paragraph-length bio that we had on file. The one we faxed to event planners who booked George as a participant, the very same bio that George had heard so often before: "Please welcome: *George Stephanopoulos!*" Yes, I could do that.

But no way. I would never create something "phoned in," something done while sitting on the Metro next to a stranger. I would write something that allowed for personal homage, that showed how grateful I continued to be for this position.

I wondered if others would envy this honor, especially the guys working in 415 who spent most of their internship in our correspondence office and not in the West Wing. Guys, and one girl, who were all older than I. Interns who at this point were probably sick to death of my pushing them to produce. These were easy relationships to strain. If I'd been stuck in that distant office most of the day and another student, younger than I was, came around like the overseer, I might not like her that much either.

I thanked Heather again and went back to work.

WE WORE WHITE GLOVES. All of us. Girls from across the country, chosen to come to Washington for Girls Nation—a true honor for each rising high school senior. We were at the White House—well, technically next door in the Old Executive Office Building. Up in Room 450. We sat, buzzing, waiting.

In 1991, I attended Michigan's Girls State program, sponsored by the American Legion Auxiliary. My high school nominated me, and I spent a summer week at Central Michigan University with other young aspiring public servants as we debated, collaborated, and ran for faux offices. I did not win any of the faux offices. Instead, I won something more tangible: a trip to Washington, D.C., to attend Girls Nation.

Remember that picture of a baby-faced President Clinton shaking President John F. Kennedy's hand in the Rose Garden? Clinton had gone to Washington as part of the Boys Nation program. After first attending the Arkansas Boys State program, he was chosen to represent his home state at the national gathering of that year's outstanding participants.

My peers floored me when they called my name in that college auditorium. This wasn't some schoolwide popularity contest.

These were serious young women, all of whom shared a passion for achievement. They were my peers, and they had voted for me to have this honor. I walked down the amphitheater stairs amid loud applause and cheers, and in that moment, I knew I could do this: I could have a future in elected politics. I stood on stage smiling, grateful, in front of rows of young women of all colors but mostly white. I felt the waves of energy—and I felt warned. A crowd's energy was just that, energy. It wasn't as nourishing as I expected it would be. In that moment, I knew that something I sought in life would never be found in the adoration of crowds. I remained on the stage as they played James Taylor's cover of "You've Got a Friend." Girls swayed and hugged, moved by having spent a week with the like-minded, feeling that energy that comes from successful collaboration. Sure, I can pull this life off, I thought. But did I want to? I did not know. We demanded the strangest morality from our politicians and a certain lockstep from those who were not yet the leader. Even then, I sensed that those strictures might make me chafe.

Like tight gloves. White gloves, to be exact. Girls Nation provided a list of required articles that included this formal wear. Michigan's Girls State had not asked for them. I vaguely knew that they were worn for formal events. The only white gloves I'd ever owned were opera-length satiny ones that had gone with my recent prom dress. I knew Girls Nation hadn't meant those.

I showed my mom the instructions. "Don't worry," she said, "we'll buy you a pair." Off to Hudson's we went, to the glove counter, asking the saleswoman to retrieve a pair from within a glass case. Ah, those. My mother explained that cotton was for summer, and they were for public outings where you're supposed to look like a young lady. We laughed over this, for they seemed such a throwback. I lived among middle- and upper-middle-class kids, and I'd never seen a girl wear white gloves. Maybe because I never went to teas or cotillions or any sort of etiquette classes, which they quite possibly did on their own time, without me, in some sort of a white-girl world they kept secret from me. Yet I didn't think so. I doubted I missed out on anything by being "white gloveless." But if Girls Nation asked, that's what we'd do. My mother paid and we left.

I could not yet appreciate the beauty or complexity of glove ritual. I saw these gloves only as deceitful, as pretty little constraints. You want me to pretend that I'm dainty? Cut to me rowing in an eight on the freshman crew team, sliding under the Potomac bridges as the sun began to rise. That's what I did all year long. Six a.m., six days a week. *Hands, body, slide, catch, pull! Hands, body, slide!*

"Don't eat or drink with them on," the etiquette book instructed. Might as well say, "Don't do anything in them but smile."

Though I loved to smile, I was never good at the implicit instruction: Shut up. Keep to safe topics. Don't rock the boat. Advice that seemed to miss the point of my life up until that time: writing and speaking out. The fact was, White House work thrilled me but I still loved writing, a love I had tried to back-burner by choosing a D.C. university over a writing program at a cloistered liberal arts college. Government or writing, government or writing, I'd thought to myself back in high school, lifting my palms up and down as if weighing fruit. At seventeen, I thought I had to choose. Government, I picked. I could always write, I reasoned. Let me learn how to both express myself and earn a living: a promise public affairs could make, but writing did not. I quickly discovered that GWU had a fantastic English department, and by my second semester I enrolled in Judith Harris's poetry workshop. When class time came around, down the elevator I went, out the lobby of Thurston Hall with a new poem in hand. But instead of turning in the White House direction, as I did when I volunteered afternoons in the Office of Media Affairs, I walked toward campus, toward my workshop in Stuart Hall, each red-bricked step a pleasure, the spring sunlight clear all around and trees full of new leaves above. And inside me a satisfaction so pure that it made the edges of my smile curl up, even though I must have looked strange, a girl delighted all by herself on Twenty-first

Street. So what? I thought. I had just finished a poem I knew could please my professor as much as it pleased me.

That semester, I wrote about my father, my mother, and their inner-city struggles. I also wrote an ode to my grandmother's hands. Our hands looked so much alike, even though I had never picked Carolina wildflowers as a girl or held a preemie baby in my arms as a nurse in a hospital unit. My grandmother may have been born in Winston-Salem, but she migrated north and made a life for herself in Motor City Detroit. What I didn't realize then was that, given her sense of etiquette, she would surely have worn white gloves with pride.

I preferred my hands uncovered, but I followed the Girls Nation instructions: whenever we left the host college grounds for field trips, we wore gloves. I felt better knowing that only a few of the participants had owned pairs before arrival. "White gloves," we'd say, shaking our heads, for here we were, learning to change the world, and they were still trying to protect us from it.

In our floral dresses and summer suits, we embarked on our tour of the White House. At one point, a guard unclicked the red rope from the stanchion and we passed through into the private residence, where we were walked outside, past the West Wing, then up to the fourth floor of the Old Executive Office Building. In 1963, Bill Clinton had met the president during their Boys Nation White House tour, but there would be no such magic for us. Neither President George H. W. Bush nor Mrs. Bush met our group, and to be honest, that was just fine with me, given my politics.

A young woman with dark brown hair and pearls gave us packets full of White House history and spoke to us briefly about her work and the White House in general. The facts I've long since forgotten. Instead, I remember her tone. The young woman sounded a bit aloof, not quite connected with us. If I remember correctly, she said she worked in the East Wing, which served as headquarters for the First Lady and her social planning, the traditional women's work of the White House before Mrs. Clinton came and took on health care and a second-floor West Wing office. Back when it was verboten for staff women to wear pants to work.

I tried to decode the distance in her voice. Was she self-conscious about speaking in front of all these young women, fearing that their gazes might see through to a place that she preferred remain unseen? I did not know. The speaker conveyed *something*, but I was too inexperienced to know what it was. Maybe it was a class thing? A voice that said, I come from Vassar or Dartmouth and I'd rather be at the Cape right now? I could decipher class codes back home, distinguishing the pretensions of girls from the old-money Grosse Pointes versus the new-money towns of Oakland County, even between girls' attitudes at the two high schools in my hometown of Troy. But out east, out here, this was all fresh, full of new languages to interpret if I wanted to understand what a person was really telling me when she spoke.

A girl raised her gloved hand and asked about internships. "Yes, you may apply for internships, of course," the speaker said. "Information is included in your packet." I remembered the exchange because I was struck by how abstract the idea seemed. I barely knew what an internship was. Now, to consider working in the White House . . . did I belong in an office with a woman like that? The idea felt so otherworldly, as if conceived for other kinds of students, the out-east kids, the "fortunate" ones Creedence Clearwater Revival sang about.

Now I was there for real, sitting in the West Wing. My fingers remained uncovered, free to shake people's hands and feel their palms against mine, just as they should be in the presidential receiving line, according to the etiquette book. Just as you would if you were holding a pencil, ready to brainstorm your speech, the one for George, the one you were going to give in front of all of your peers right there in Room 450 two years later.

* * *

SPEECH DAY ARRIVED. Heather and I walked to Room 450 early to put water on the lectern and make sure everything was okay. "Are you nervous?" Heather asked. "Yes!" I said. "Don't be," she advised me. "You're going to do great." I appreciated the vote of confidence. I sipped my own water, willing myself not to spill any on myself or my speech, not to do anything to sabotage this moment.

Interns packed the auditorium to the last row as they talked, ate their lunches, and waited for George to arrive. They filled the space with buzzy happiness. The other correspondence interns walked over to us as soon as we entered, and we shared a collegial moment.

Karen, the director of the internship program, came over and spoke with Heather. She greeted me warmly, too, even though I was not officially part of the internship program and I was introducing George at an intern event. She seemed unbothered by the fact that I had found my way to the West Wing through channels other than hers. I was glad.

George entered. All eyes turned his way as he walked toward us in front of the stage. Quick greetings, smiles. That was Karen's cue to begin the event.

This wasn't like debate tournaments or Model United Nations. This wasn't like Girls State, either, standing in front of everyone, accepting an award. This was different. This was a level of performance I wasn't used to, a desire from within to really deliver. Nervousness turned into stomach-pounding fuel.

Karen announced that George would be introduced by Stacy Parker. As my peers applauded, I stood up and walked to the lectern, placed my prepared speech on the slanted wood, and began:

> As White House interns, we have all had moments that have given us insight on the presidency, the media, and the American people. And while there are times the three of them seem to resemble a Shriners' circus, we know that our experiences

here have led to a greater understanding of the political process and the people of our nation.

This summer, many of my touching moments were spent opening the mail of Mr. George Stephanopoulos. He is the senior adviser whom we all know has the willing ear of our president and the ability to "make things happen." While part of this summer was spent learning how difficult change is for some of the American populace, much of it was spent learning about the hope that President Clinton spoke of so often during the campaign.

For many of the American people, especially our generation, George Stephanopoulos personifies this newfound hope in our system. All one needs to do is look at the Post-it board in Room 415, the room where George's mail is answered, to see the fruits and offerings of peoples' wishes and dreams:

There is the picture of the brown-eyed altar boy sent in by a mother who prays daily that her son will grow up to be "just like George."

There are the news clippings in Greek, French, and Korean sent in by admirers who have felt the need to connect with the man who has touched them through their newsprint and television screens.

There are the pictures of eligible young ladies sent in by aunts who believe that their nieces would make perfect companions for such a rising star.

And of course, there is the black plastic comb sent in by a grandmother who complained that his hair covered his beautiful face when on the morning news shows.

But my favorite is the sign. Printed on brown paper with bright blue, red, and green markers are the words: "Thank you for doing something." This sign was sent in by a student who made it to wave at President Clinton when he visited New York after the election. She sent it to George because as she said in her accompanying letter: "you deserve it."

And yes, George Stephanopoulos deserves it. He has spent almost his entire professional life serving the public and serving it well.

Currently, he is the senior adviser to the president for policy and strategy. During the Clinton/Gore campaign, George served as director of communications and oversaw polling, policy, scheduling, press relations, and media operations.

Before joining Governor Clinton's campaign, he was executive floor manager to House Majority Leader Richard A. Gephardt. In 1988, he was deputy communications director for the Dukakis/Bentsen campaign. Before that, he was administrative assistant for Representative Edward Feighan of Ohio.

George received his master of theology at Balliol College, Oxford University—where he studied as a Rhodes Scholar. He obtained his BA in 1982 from Columbia University and graduated summa cum laude in political science.

For those of us who spent our formative years trying to make sense of the deceit and narcissism of the Reagan/Bush era, George Stephanopoulos inspires confidence and hope. President Clinton and George Stephanopoulos reaffirm my faith that with a good head, a good heart and enough political savvy, one can do good things in Washington. Would you please welcome Mr. George Stephanopoulos . . .

Applause, applause, deep applause for the man as George shook my hand and took his place behind the lectern. "What Stacy forgot to mention," he said, "were all the eligible young men" (who sent in photos). The audience laughed. They had also laughed knowingly at my "deceit and narcissism" line. I sat next to Heather and she squeezed my hand, whispered that I had done a fantastic job. Afterward, I was congratulated by other interns in the room, only some of whom I knew. I was on top of the world.

Back in the office, Heather told me that George had been impressed: "She really worked on that speech," he said. Now I could relax. I felt as if I had cleared the bar with inches to spare.

LATER THAT AFTERNOON, I walked up to 415 to retrieve some letters for George's signature and to bask some more in the glow. Up the wide, winding stone stairs to the top of the castle, to the kind of room a princess would have been held captive inside by an evil sister. I hoped the guys would have more nice things to say.

The conversation stopped once I entered. *You guys doing OK? Sure. What's going on? Nothing. You did a nice job. Thank you.*

One of the guys handed me the bin they knew I needed, and as there seemed to be nothing more to say, I turned around and left.

I don't know why I had expected more. This was exactly what the good soldier inside had wanted: efficiency. I hadn't been there three minutes before everyone was back to work.

GOAL-DRIVEN. In the eight, we raced to the finish line. We did not chat one another up in the boat—breathing, oxygen, that's what our mouths cared about when we pushed ourselves hard on the water. Maybe on the jog to the boathouse or in the Thurston basement cafeteria for breakfast afterward we'd talk and laugh, even though we had to force ourselves off the one topic we had in common. *Come on, guys, just for one breakfast we're not talking about crew!* We'd all agree. Then, after ten minutes, we couldn't help ourselves. Back to crew. Then off to class.

Psychologically, once I entered the Seventeenth and G Street gate I was in the eight again. I passed through the magnetometers and I raced. I kept a steady gait, kept a smile on my face despite the mounds of mail, despite knowing that Heather wanted to be caught up by summer's end. So when I entered 415 OEOB, I greeted everyone, asked how everyone was, then booted up the computer. I opened letters. I coded letters. I typed responses. Just as all the other interns did. Conversations started, but I tried not to continue them, forgoing the chance to get to know the others better. My unease with the guys might have been a major contributing factor. Still, the backlog situation felt like a crisis. Failure was not an option. Oar in the water, push with the legs, I thought. Oar in the water, push with the legs.

But after the brown-bag lunch, my old high school sociology teacher Mr. Lage was on my mind. "It's not what you teach but how you treat people that they'll remember." He'd said that to us in class, but I connected that wisdom to the time he had singled me out for mentorship.

Mr. Lage kept his classroom seats in a big circle, my only high school teacher to do so, and he came and sat in the desk next to mine. He explained to me that I was being rude in class. I was mortified. He continued, saying that maybe I didn't realize it, but I was openly disrespectful when I did homework for other classes in front of my teachers. That I should think about the message that sent. He said this to me so compassionately that I immediately knew how I looked: selfish, so laser-focused on a goal that I forgot the human feelings involved. My body spoke without my mouth saying one word: *What you are doing up there is not as important as this work in front of me, work that I should have finished at home last night.* Sometimes students had no choice, we both knew, but he wanted me to understand the unintended consequences of my actions.

As I waited for the OEOB elevator, my hands hurt from the mail bin I was carrying and I set it down. I interacted with staffers who were all smiles for George or Heather, but once they looked at me, a mere intern, their faces went blank. I never wanted to be that person. Sometimes Heather could be short with me, but I always knew she cared about my well-being. Could the other interns say the same about me? I promised myself I would do better. Maybe I couldn't improve much with these guys, but with the second batch of summer interns I would try harder to develop rapport, to know that our time here was about more than processing George's mail.

We could perform without being jerks, without the empty etiquette of false smiles and white gloves. I knew this in my heart. But if I needed reminders, I could look to George, to Heather, who showed me in deed every day.

6

This Kind of Intimacy

HEATHER SPENT THE AFTERNOON ON JENNIFER'S BED. THAT'S what she told me the day she returned from her three-hour lunch spent with Jennifer Grey, George's current girlfriend.

I knew Jennifer as the character Jeanie Bueller, Matthew Broderick's sister in *Ferris Bueller's Day Off,* for I grew up in the John Hughes generation, staring at the screen and dreaming myself into his movies, including this beloved one about what happens when a high school senior decides to take his best friend and his girlfriend out for a day on the town. I loved the scene in the police station where a gorgeous hoodlum, played by Charlie Sheen, advised the resentful Grey to lay off her brother—and the makeup. "Shauna . . . Shauna . . ." the doo-wop group sang as she exited down the stairwell, flustered by his attentions.

Now she was starring in *The Twilight of the Golds* at the Kennedy Center. And dating my boss.

"Tell me more!" I demanded of Heather. George was gone and Heather and I were alone in the anteroom and I did not care that I was exhausted from "playing Heather" for so long, having covered the phones as they rang off the hook. I wanted to know more about her afternoon with the movie star. Heather never took lunches, much less three-hour lunches outside the White House gates.

Was Jennifer using Heather as a "George decoder," trying to get a read on puzzling aspects of his behavior? Jennifer had once called the office and asked me about "The Mister's" mood. This had made me laugh inside. In fact, everything about her query was funny to me, from the idea of sharing in her lover's term of endearment to the notion that George had more than one visible mood. He stayed cool with us, never confusing the roles. He did his job, and we interns did ours. I thought of him as a calm ocean, with all the action and mystery living several leagues below. Places, I thought, that only a girlfriend could go.

"We just sat and talked," Heather said, slipping back into her black chair. "Girl talk," she added. Then she picked up the phone and began to return her calls. *Please, Heather, tell me more*, I wanted to beg, but I knew that Heather considered the discussion closed.

Girl talk with a movie star who was dating your boss after a blind date you'd helped set up: that was one way to spend the afternoon! I heard other assistants snipe that Heather was in love with George and that's why she was so protective. But I dismissed those accusations as easy, bitchy things to say about a woman working for a man. I felt protective of George, too, but I was not in love with him. Whatever I felt had a different name, and I assumed the same for Heather, too. Given how attractive Heather was, I figured that if she wanted George, she would have had him already.

I went back to my beta role in the intern seat and wondered what Heather thought of this new life, six months in. Men and women she'd shared BBQ with at Doe's Eat Place in Little Rock were suddenly the new cast of characters in the national play, their lives in a white-hot glare. From zero to celebrity just like that.

Heather stage-managed both the man and the celebrity—this public character who was not just a character but a new archetype that had entered the cultural imagination. The "George" role was now the competent, earnest young male aide, wearing gold-rimmed glasses just below his furrowed brow. This was the character that would be inhabited by Michael J. Fox in the film *The American President*, then in the television comedy *Spin City*, and later by Rob Lowe in the acclaimed television drama *The West Wing*. The following year, a producer from *Friends* would call, trying to snag George for a cameo. George said no, but *Friends* still revolved an episode plot around him: the girls received George's pizza delivery by mistake, allowing the most famous women on TV to moon over him in prime time.

George the man seemed unfazed. Heather took all of this in her cool Britannia stride, too. But my imagination was on fire, trying to see life through George's eyes, to understand what this meant for him. Maybe it was like going to the Macy's Thanksgiving Day Parade and seeing a rubber version of yourself blown up and "walking" with the help of a dozen attendants, this version of you more than ten stories tall, knowing that your celebrity was just that, something outside you, something as big and as vulnerable as giant balloons.

LETTERS. OPEN THEM. READ THEM. Fill out the code sheet we had created, put the letters on the pile. Respond. The same old routine until I reached in and pulled up a love letter from George's ex-girlfriend.

No one had ever spoken directly to me about his ex, but there was so much I knew that no one had said directly to me, as I looked with my eyes and not my mouth and gleaned from nearby conversations and *In the Loop* mentions and all else I picked up by, as the band XTC would put it, having five senses working overtime.

The letter was already opened. Someone had read this, I assumed George. It must have been George. But if it was lying in a white U.S. mail bin here in OEOB 415, we interns were supposed to respond to the letter, then send it on to the Office of Records Management. Anything addressed to someone at the White House was supposed to be archived. But who had read this? If it was George, did he mean to have a letter from his former girlfriend processed by us, then sent to Mr. Terry Good's shop for safekeeping?

I held my breath as I held the letter. Should I read it? My God, this was his *ex* we were talking about. I wondered if Jennifer Grey knew they were still in contact. What had his ex-girlfriend written?

The other interns were busy reading other correspondence, and I held the ex's letter in my hand. Before a human does anything there is that hesitation, that small moment, if we are conscious enough to notice and take heed, when we can stop ourselves from pulling the trigger. The free-will moment. I felt it.

I opened the letter anyway. I read it until I got to the place where she started talking about their past love life and how good it used to be. I read this as I would any gossip—hungrily. And though I felt guilty for what was undoubtedly an invasion of privacy on my part, I felt happier knowing that George made her happy. To me, that vouched for him more than the glossy profiles, or the adoration of a million fans.

I snapped to. I told everyone I would be back, that I needed to run something over to Heather.

I mouthed the ex's name as I gave Heather the letter. She looked at me, then took it from my hands. "Okay," she said. That's the last I saw of it. I felt for the woman, and for George, now living these plate-glass lives where something as private as a love letter was private no more.

At least if you mailed it to the White House.

* * *

IN THE PICTURE I snapped, you could see Emma with her wavy auburn hair and sweet smile leaning next to George's car, the red Honda CRX with the Clinton/Gore sticker parked on West Exec. Emma had a goofy *hee hee* smile on her face, because we took this on the sneak—laughing in that deep diaphragm way that I did so much with my friends at that age when something was so funny you thought you just might die. This felt hilarious because we were standing between the OEOB and the West Wing, knowing that if I had caught other girls doing this, I might have stared daggers, wondering what in the world they thought they were up to. But this was *us*. Emma served on the second-session team devoted to George's correspondence. Emma and I worked hard together answering George's mail. We clicked, too. I did better with the second-session interns, for I did not overstress my role as a manager and I learned that by giving rein while remaining attentive, we could maximize both productivity and bonhomie in the office. Emma and I bonded over our shared sense of mission, believing that each fine response written for "GRS" (George Robert Stephanopoulos) cultivated support for the administration. We were doing our part. And we were helping George, the guy we respected and admired.

They sure look like groupies, don't they . . .

Oh no, the dreaded word! Quick, quick—to be caught taking pictures by his car would put us square with those girls (and boys) in the white mail bins. As if we were fans, engaged in fantasies untethered to reality. Well, let me tell you, Emma and I had rapport—that made us different, right? We knew that we were neither his confidantes nor his friends, that we were there to help, but we were not like girls who watched him give press conferences and thought that just because they watched George on TV from the comfort of their bedrooms that they somehow shared genuine intimacy.

Click! Keep walking as if nothing happened. As if nobody saw!

I wondered if from then on, whenever Emma saw a red CRX pass, would she look for the sticker? For that's how I knew it was George who passed me on M Street when I walked into Washington Sports Club, the same gym he worked out at every day after work. Thank God I had joined before I started working in his office. I was a regular there, too. Those days, I did a maintenance routine of twenty minutes on the recumbent bike or the StairMaster, then moved on to light weights.

George spent forty minutes on the StairMaster—which back then seemed like forever—and did so at a high level of difficulty, reading the newspapers he hadn't finished earlier in the day. Yes, the same gaze of mine that lingered on blossoms or on burned-out buildings back home, looking for telling details, would zero in on George whenever he entered the room. I'm sure I wasn't the only one. Power attracted attention. People looked. However, nobody bothered George much on the gym floor. If George and I saw each other, he smiled and said hello and I felt special for the rest of the time, for I *knew* George and all these others just wished they did.

One time, I was sitting on a machine in between sets and my mind had gone blank. George passed me and stopped. "You look like you're in deep thought," he said. I shook my head and said, "No, not at all." I honestly had nothing on my mind. I spent the rest of my workout grateful that the one time I knew George had focused on me, I had looked thoughtful!

Because Lord knows we interns focused on him, and usually assumed that if he was quiet, he was in deep thought. Not because we had crushes—what we had was something different. An affection, maybe. An affection that was crushlike because our desire to care for and protect him was not born of the give-and-take between George and ourselves. We started our jobs with those feelings. But the feelings grew the more we worked with him, maybe due to the implied hope: I do well when you do well. Protect the leader, and I am protected, too.

I did not think about such bargains at the time. Gym encounters notwithstanding, I knew that whatever intimacy I felt with George was with his hologram projection. The kind of intimacy that was no intimacy at all, or at least not the kind shared by lovers, the kind I wanted, the one when you opened yourselves to each other.

"Is he still dating Jennifer Grey?" Emma asked as she let her camera hang from her wrist. I nodded. I told her how Heather had spent a three-hour lunch at Jennifer's place, lying on her bed and talking girl talk.

"What's the latest with the stalker?" she asked.

I raised my eyebrows. If George needed more evidence that his life had changed, there it was: a real stalker. A woman followed George sometimes. A woman Heather knew about. Knew she called. Knew to be on the lookout for. Heather did not tell me much, but there was a dangerous woman out there consuming all that celebrity, believing that George was meant for her and her alone, trying to make contact, trying to break through.

"I think it's under control," I said, and Emma nodded.

Stalker. We used the word so easily, all of us. No one I knew fit the criminal profile, but all of my girlfriends and I had done things that fell somewhere on the stalker continuum. Like the way we used to drive past ex-boyfriends' houses to see if the ex was home, and if so, with whom. We could spend a whole night drinking vodka and pink lemonade and listening to the love songs station, then drive by, cursing his name but wanting him desperately. We didn't think we were psycho. We felt the "psycho" line was crossed if you attempted to make contact, if you walked up and knocked on the door and tried to talk to someone who wasn't expecting you, who might not welcome you. To do *that* was psycho. To do that meant you were really a "stalker"—you'd crossed over from common human jealousy into something else.

Emma and I climbed up the stairs from West Exec and entered the West Wing through the lobby. This time, Emma snapped a photo of me. People toured here all the time, and taking pictures was normal. Brent, the lobby agent, even offered to take one for us. We smiled. *Click.* Then Emma took one of me at Brent's side before we went and picked up a new batch of incoming mail.

THERE WAS NO CALL from Charles. He made me achy in my throat, in my ribs, when he ignored me for days on end. My pride was strong. I did not call him much, taking the hint that if he wasn't calling, he wasn't that interested. But by midsummer, I calculated that I had slept with Charles more often than with any other lover. I tried to avoid casual sex, given my fear of pregnancy, which, at nineteen, would feel tantamount to disappointing all the adults who had ever invested themselves in my future. Worse than pregnancy, though, I feared contracting an incurable disease, such as herpes or HIV, that I assumed would make me a woman whom no man would ever touch again. Life without touch would be punishment by solitary confinement, only in plain sight. Like an unwatered plant, I'd shrivel up and die. I've since come to know HIV-positive people who've created strong, loving bonds after diagnosis. But back then, I considered HIV the most tragic of fates: we needed intimacy to live, but that same intimacy could kill us, too.

I set aside these worries with Charles. Being with him made me happy. Just to be in his car at night as he turned onto Virginia Avenue, the soft lurch as he shifted gears. He would speed his compact cabriolet past the Lincoln Memorial, across the Memorial Bridge, over the black water below to the Virginia side, driving up and over that gently undulating land, past the Pentagon, past the soldiers on duty, listening in, maybe, with their reconnaissance toys. Listening to us say nothing. Just feeling the engine and the air rushing past the rolled-down windows and Alabama's "Dixieland Delight" on the mix tape. My mother had played country music all through my childhood, and I wondered if his had too or if he had learned to love it at Vanderbilt University, from which he had recently graduated. It was a conservative place, I worried, but his parents must have been so progressive, given that they were professors at a New York liberal arts college; I doubted that they would openly dissuade their son from seeing me, or be afraid that my color might slow his career trajectory. A boy I had dated in high school—a boy with whom I thought I had a future—had succumbed to such reasoning, and the memory of his rejection still hurt.

For that was my dream: that we were not just fucking and putting ourselves at risk but laying the groundwork for something big. That we could be the kind of couple journalists wrote profiles about. Formidable. He worked in congressional affairs for a powerful law firm. For a boy like him, the sky was the limit. I thought it was for me, too.

That night, Charles played with the hem of my denim miniskirt. We rode low and fast, and his touch tickled in all the right places. I anticipated lying back in his bed, letting him lead, knowing that in his bedroom, he never disappointed.

Then there were other nights. The majority of nights. The nights he never called.

My thirty-year-old friend Allen told me that when I got older, lovers wouldn't come and go as quickly as they did now. I hoped he was right. Charles wasn't gone, but he felt gone, and I knew I needed to accept that if this were the road to love, it would feel, and be, different.

HEATHER LIVED in an apartment building near Thurston Hall in Foggy Bottom, almost equidistant from my old freshman dorm and the White House. I never knew this until the July afternoon when she said, "Stacy, would you like to come over after work, just for a bit? I would like to talk to you and do it outside of this"—her eyes shifting from wall to wall in the anteroom of our world.

A few hours later we exited Seventeenth and G Street together, into the evening that had barely cooled. We walked the short blocks to her one-bedroom where I imagined Heather Beckel could be herself, no longer squeezed into her White House role as George Stephanopoulos's executive assistant.

Walking up E Street, she said, "Stacy, this fall, I have permission to hire a staff assistant. I know you're devoted to your studies, but I feel compelled to tell you about this development, in case you might want to take this job. You know the place. You would be perfect. I'm not convinced that you should delay your university work to do this. But you are my first choice. You really are."

"Thank you, Heather," I said.

"If you have the desire," she continued, "let me know. But I don't expect it, okay?"

What Heather was proposing seemed incredible to me: to be actual White House staff at the age of nineteen. I loved school, but Heather was providing an opportunity that could not be summarily dismissed: not only to work at such a prestigious address for people I believed in but also, for the first time in my short life, to make a living wage by doing so. I had always wanted to be George. But should I put off school to take care of George?

"Heather, don't count me out yet," I said.

"Wow, you'd really think about it?" she asked.

"Of course, Heather. It's for you. And George. That's the least I can do."

LAMPLIGHT WARMED THE CORNERS, one by one, as Heather walked around the dark-curtained space of her apartment sitting room, revealing low pieces of retro furniture against the walls. In so many

D.C. offices, you found prominently displayed framed pictures of the person whose office it was shaking hands—or better yet, arm in arm—with important politicians and Hollywood stars. At work, neither Heather nor George had such an "ego wall." Each received pictures taken by White House photographers and occasionally displayed them, but there was no showy effort to prove what their history and real estate proved in spades—that they both had intimate access to power.

But look at what hung on Heather's home walls: A still of George calling President Bush on the Larry King show. A childhood snapshot of President Clinton. A sunny portrait of the president, the vice president, and their wives, taken on the postconvention bus tour. Heather as a blonde. Heather with her handsome brother. A formal class portrait from her childhood school in England.

I lingered over the young Heather in black and white.

What was it about Heather? I felt such warm mother love from my own mother that I was hard-pressed to understand why I cleaved to Heather, why I so quickly identified with her. Maybe because she was my guide in this new jungle. Or maybe because she showed such care for me and my future and the symbiosis felt was actually real.

"If you decide to stay in school, which I think you should," Heather said as I joined her on her Victorian couch, "I still want you to intern for us. I will need you to train the fall interns, okay? Show them everything you all accomplished in 415. You'll still help manage the correspondence."

She knew. For twenty minutes I had considered the possibility, but college was just too important. I put my tea down and hugged her.

George had broken ground with his young self, working the halls of power. But that was George. I was just an intern. And for now, that was fine. "Intern" was code for a student with a future. A big, bright professional future. Keep up the good work, and I could be the next George. Not the next George's assistant.

I left Heather's place feeling chosen. Nothing could mar this internship experience.

JULY 21, 1993, and I was sitting in my chair feeling sick. I could hear Heather crying. Not loudly, for I could tell she was trying to silence herself, but when I turned, I saw her pink eyes and nose and I knew it was because of the news: the night before, White House Deputy Counsel Vince Foster had shot himself in Fort Marcy Park in Virginia. I did not know Vince Foster. I had seen him only once or twice in the hallways. He worked upstairs. He did not visit or call George regularly. But the shock felt so thick, as if all of us, no matter our connection to him, had been punched in the stomach. Suicide. The last time I had encountered suicide was in high school, a boy named Jason, a boy I did not know but whose death deeply affected friends of mine, including Jeffrey, who still has a memorial tattoo above his ankle. Such a shock makes you sad for the family's loss, for your own loss. But it also makes you reevaluate your life. The president's opponents tried to block every move he made, and the tempo could seem impossibly hectic in the West Wing, but this kind of life felt so valiant, so exciting, that I had never guessed that others were racked with horrible pain. Heather and I sat in George's office and watched on TV as the new communications director, Mark Gearan, spoke about Foster. Both of us cried, quickly wiping away the tears.

Was it the pressure that had made Foster do it? He came from Arkansas, from the Rose Law Firm, a close colleague and friend of Mrs. Clinton. I later learned he had been tasked to help manage the escalating "Travelgate" situation: the charges that Mrs. Clinton, and others, had summarily fired the White House Travel Office head, Billy Dale, and the rest of the staff, and had their FBI files rummaged through for the purpose of finding damaging information they could use as an excuse to replace them with folks they knew from Arkansas. Billy Dale was tried for embezzlement and found not guilty. The other six staffers were fully exonerated. Eventually, Mrs. Clinton was cleared of wrongdoing by Independent Counsel Kenneth Starr. But in 2000, Independent Counsel Robert Ray said that First Lady Hillary R. Clinton "ultimately influenced" the proceedings. Clearly Vince Foster had a difficult situation on his hands.

Carl Bernstein, in his 2007 book *A Woman in Charge: The Life of Hillary Rodham Clinton*, wrote of how the married Foster and Hillary Clinton had been very close, that many in Arkansas thought they were lovers—if not physically involved, then deeply intimate on a mental and emotional level. True or not, Foster apparently felt deep anguish as congressional committees deepened their investigations into the firings and the role Mrs. Clinton had or had not played in them. He seemed to feel he was not doing a good enough job to protect her. He also suffered from depression.

At the time, I knew none of this. I just knew that one of our senior staff had committed suicide. Not just "staff" but a genteel, well-respected man who had moved from Little Rock to help his dear friend who headed up the Health Care Task Force in the administration that had promised to provide everyone with health care.

Soon rumors of murder and conspiracy poisoned the air, spurred on by the fact that files had quickly been removed from Foster's office. "He knew too much" snowballed into "Hillary ordered him dead!" Rumor artists used the clumsiest conspiracy-theory logic to blame the Clintons for his death, as well as the deaths of journalists supposedly posed to reveal all. Patrick Matrisciana did so in his 1994 film *The Clinton Chronicles*—a video sold through infomercials and featuring an appearance by the politically influential Reverend Jerry Falwell, a demagogue of a conservative who would later be quoted as blaming American gays, feminists, and abortionists for the September 11, 2001, terrorist attacks.

Mrs. Clinton later branded these storytellers as members of a "vast right-wing conspiracy"—the writers, lawyers, publishers, and funders who worked together to get their Clinton accusations into the bloodstream of the body politic. The billionaire Richard Mellon Scaife funded the "Arkansas Project," supporting journalists and investigators who dug up or made up dirt on the Clintons' past, including theories about Vince Foster's suicide, the Whitewater investment deals gone wrong, and the Arkansas state troopers who had purportedly procured women for the then governor.

The "conspiracy's" efforts bore fruit. Stories about Troopergate printed in Scaife's *American Spectator* introduced us to "Paula," who had accepted then-Governor Clinton's invitation to meet him in an Excelsior Hotel suite. This "Paula" was Paula Corbin Jones, who in May 1994 would file a civil suit against President Clinton for sexual harassment. A suit that eventually would subpoena the then-unknown Monica Lewinsky as a witness, for, having been rumored to have had an affair with Bill Clinton, she could verify that yes, the president did involve himself sexually with female underlings. Jones's lawyers served Lewinsky with a subpoena, and with that came the accusations that the president coached Monica on her answers. These accusations led to the obstruction of justice charges at the heart of the 1998 impeachment case against Clinton, the second impeachment ever of a sitting U.S. president.

But back then, in 1993, I closed my ears, thinking such charges were just hateful attempts to diminish a man whose opponents hadn't beaten him fair and square in the election. When I later learned of Scaife's bankroll, I dug my heels deeper into this position. If I heard whispers of womanizing or of Hillary throwing books at her agents, my brain immediately classified the stories as political dirty tricks. If the accuser stepped to the microphone and she looked to be surrounded by partisan Republicans or like someone who hoped to make money off the story, I didn't hear the content of the charges at all. I immediately dismissed the accusers as opportunists and kept on with my work.

EVENTUALLY WE STOPPED CRYING, at least in the office. I knew I was lucky to be an intern. Our lives were coddled. Become staff, I thought, and your life became exposed. Just as you targeted opponents in the campaign, you could become a target yourself.

Until then, I had never imagined what it was like to have critics hound you in the media. I never thought you could feel so cornered that you might look at a pistol and think that was your only way out.

When I read profiles of staffers in newspapers and magazines, I started looking for mentions of reality, of a sense that this wasn't just some magical glamour land but real people with real lives who had no power to suspend natural laws of the universe. The president and his staff had power, yes, but the weight of it, and the fight to keep it, took a toll on the body and psyche.

But if anyone asked me if George was still dating Jennifer Grey, I said yes, yes indeed. Our craving for knowledge of one another's personal lives, especially those of the famous and powerful around us, was not going away anytime soon.

7

Trespassers

Washington, D.C.
August 1993

"I'VE HAD MENTORSHIP GO WRONG," SAID TRISTAIN* OVER THE LIP of his yard glass. He raised the beer with both hands and drank, his dark hair lustrous in the lamplight.

We sat in Tiber Creek Pub a few hours into the Friday-night happy hour, the place packed with young adults creating a roar louder than the *you can't always get what you want* piped in from above. Four of us sat at a small four-top: Tristain and Chito, who interned in the White House news analysis office, along with me and Wendy, an old friend from Troy High School who interned on the Hill. I drank from my own half yard of beer but carefully, now that my damp shirt was drying from what had already slipped down my chest.

"My favorite professor," he said. "We were really close. But then he came on to me."

---

*   This name has been changed.

We exchanged alarmed looks. Tristain remained silent, as if the emotion of the moment had rushed back to him.

"It used to be that we could talk about anything," he said. "I thought, so this is what they meant by the mentor/protégé connection."

Oh God, I thought. I imagined the flush of realization when Tristain felt the professor reach for him, the slow-motion moment when the professor killed all ambiguity. Gone was the dream where Tristain could want one thing and the professor another and both could pretend those things were the same.

I have to admit that I felt for his professor. Tristain looked like a punk-boy Cary Grant. I knew him because he worked in the office directly across from 415 OEOB; he was one of the second-session summer interns tasked with assembling newspaper clips. After first spying the twenty-one-year-old in our hallway, I spent a week gushing to Heather and Office of Media Affairs aide Keith Boykin about his dark hair, his fair skin, and the confident way he walked. This was August now, and Charles and I were no longer dating. Sometimes he called late at night, offering to pay for the taxi out to Alexandria. Sometimes I accepted. I knew that we were just hooking up, my hopes for love pretty much dashed. In the meantime, I kept pining aloud for Tristain until Keith, his boss, finally told Tristain to go talk to me.

He did. The week before, he had knocked on our 415 door and sat down on our red couch, the color of tomato soup. Or desire deferred, I thought, as he talked to Emma and me for twenty minutes and I kept my crush to myself. I thought of the star boys I'd loved before, the musicians in real live bands who thought a good time meant jamming while the girls sat and watched. Not Tristain. He seemed interested in us as individuals.

"What did you do?" asked Wendy, loud enough to be heard over the crowd. "I let him down gently," Tristain said. "I told him that I cared for him, but I didn't like him that way. *I'm not gay.*"

We laughed, but Tristain shook his head. I sensed acceptance on his part, a level of wisdom about human frailty you didn't find that often among young adults drinking at a bar.

"I guess I thought only girls had that problem," I said.

"How did he take it?" asked Wendy.

"He took it well. I guess," said Tristain. "I think our relationship took a hit, you know, but we still talk."

Three hours earlier, Tristain had walked into our 415 office, looking to do just that. Emma and I were there, restless, trying to make our correspondence quota. I had the best of the British pop group ABC in the tape deck. *Shoot that poison arrow through my he-a-a-r-rt.*

Tristain knew the words! He saw my delight as he lip-synced. Then, like a movie star, he jumped up on the red cushions of the couch, the couch that had come with the room, the nest of how many Bush administration liaisons I didn't dare imagine. He walked straight in and jumped up as if the couch were his stage, and he belted out the words, my heart melting that he, this beautiful boy with the dark hair and the light eyes, knew the words of this deeply romantic song. All I wanted to do was fall to my knees, as if bobby sox peeked from beneath my full skirt, and stare up and adore this boy.

"Come out to the bar with us!" I asked him. He said "Sure," then broke into a big grin. Emma had plans she couldn't break. So we were a foursome there, sitting at our table, with me wishing we could break into two again and Tristain would be left to me.

"How did the whole thing make you feel?" I asked.

"It's just sad," Tristain answered. "I thought he was my mentor. You think you have one kind of relationship, a father-son thing, or even an older brother and younger brother thing, and then you realize it's something else completely."

"A toast, then," I said. "To Tristain surviving his near molestation!" We laughed and reached the monstrous mouths of our test tubes of beer to one another, beginning the ritual of lifting and balancing so all the beer that settled in the bulb did not come splashing down our faces.

MENTORSHIP. Might as well be called "magic," the way teachers and motivational speakers present it at "emerging leaders" meetings and student retreats as the ultimate key to success. Mentors would be our initiators into the rituals of the professions, the teachers of rules that didn't get committed to print. No one ever mentioned that mentors could be predators, too.

I realized I had been lucky so far. If my male teachers had ever committed sins, they were the cool crimes of indifference: I can picture two math teachers in school who stood idly by as I floundered, their equations as inscrutable to me as hieroglyphics. But that's as bad as it ever got; no one ever crossed generational boundaries with me in a greedy, amoral way.

In fact, when my teachers reached out I felt only goodness. Like my high school economics teacher, Mr. Bradin. He was an unlikely superhero in his midfifties: quiet, with his white hair, pink skin, short-sleeved shirts with ties. I loved how he emoted a happy nerdiness as he taught us about Josef Stalin's five-year economic plans for the Soviet Union, exuding a certain *aren't you glad we have competition and choices in America?* gratitude without being dogmatic.

One afternoon, Mr. Bradin spent half the class talking of life under Stalin's purges, explaining how Stalin had exiled and murdered his own people in ways systematic yet capricious at once— killing whomever he thought he needed to kill in order to maintain power, or stoke the fears of those who remained. Mr. Bradin told us what it meant to hear that knock in the middle of the night. *That's it. You're dead. Not necessarily because you did anything. Maybe Stalin was purging all those whose names began with P that week, and you were Parker. Or Parkervich.*

Ha ha ha, I thought, pursing my lips when he looked my way. But then the words sank in. To be purged. To be so utterly powerless. To always be afraid.

Strangely, chewing tobacco was the rage my senior year, the country habit tried on by the boys of my suburban school. I sat there in my seat in front of the back-row seniors, each with just enough dip packed next to his gums to avoid making a bulge in his cheek. They spit into the old carpet, thinking the teachers wouldn't notice or care—conceivable, since the school would be moving to a new campus across town the following year. Still, the baseness of it bothered me. They spit, and I sat in my metal desk, overcome with a knowledge I had tried to ignore up until that moment: that any of these boys could overpower me if they wanted.

Within minutes, I had a terrible thought: this fear life wasn't hopelessly foreign, like some flip-side world revealed once the kidnapper's burlap was pulled from the head. I lived in that system already! Why? Because I was a girl. A girl in a man's world. Sure, this was 1992, and despite being athletic, five foot ten, and outspoken, the traits that should have meant, or at least telegraphed, that I was strong, I was still physically weaker than all the men I knew, given their upper-body strength versus my own. What a cruel joke: as if nature decreed that men should always have the ability to overpower me if they chose. All they needed to do was push me over. How could I ever expect to be equal, to be unafraid?

I could buy a gun, I thought. Firearms were the great equalizer, no? A solution that comforted me until I imagined myself being overpowered and having my weapon used against me.

Silently, red-eyed, I started to cry. The bell rang, and the class moved out the door. I picked up my notebooks slowly from beneath the chair and stuck them into my big army surplus backpack. Mr. Bradin came over. In a soft voice, he asked me what was wrong.

The words came tumbling out. "I am going to be vulnerable for the rest of my life just because I'm female." The tears fell, and

I wiped them away quickly. Men were stronger, and there was no getting around it.

He looked surprised, but only for a moment. I'd never opened up to Mr. Bradin before. He wasn't one of those teachers, like Mrs. Nixon-John, who was constantly approached by students who needed to be listened to by discreet, sympathetic ears. Mrs. Nixon-John, my English and creative writing teacher, had become more than a confidante: she was my mentor. The definition of kind, generous guidance. She edited my poems. She invited me to her nearby home to type on her Mac II, since I did not yet own a computer of my own. Those evenings, I sat at her family table for dinner, alongside her husband, Mike, and sometimes her grown daughters, Renée and April; I delighted in the first marinara from scratch I'd ever had, dumping spoonfuls of fresh Parmesan on top of the homemade pasta, all of this new to me and giving me deep pleasure. She also ladled praise and encouragement, but only after she marked up my pieces, giving me my first taste of strict professional attention. With Mrs. Nixon-John I learned how to take critiques, not to be defensive, to know in my heart that her slashes and arrows built up and never tore down. With her, I learned that *all* writers were edited, that it wasn't cheating or a strike against the quality of my work. With Mrs. Nixon-John's encouragement, I entered high school writing contests and even placed first. Once I started college, I sent her poems from my poetry workshops for her critique. Once I started in George's office, I took Gloria—for now I was to call her Gloria—and her daughter Renée on a tour of the West Wing. This is what I knew of the mentor/protégée relationship: female, exacting, and nurturing. Kindness to be reciprocated with gratitude and gestures along the way.

Mr. Bradin closed the door. He sat down in the seat next to mine. "Stacy," he said, "you are not doomed to be powerless just because you are female. Life is risky, period. One can never mitigate risks completely. We manage them. As in economics, in life.

We learn to discern dangers—the trick is getting out of situations before they escalate."

I listened. I nodded. But I must not have looked convinced, for he suggested that maybe I should take a self-defense class. That this could make me feel more empowered. Nodding, I thanked him for the advice and for his kindness, and I left for my next class, not believing that I had had these poignant moments with Mr. Five-Year-Plan Bradin.

I never signed up for a self-defense class. I never bought a gun either, for I didn't want the energy of a killing tool in my home or on my person. But I always kept the memory of that day when Mr. Bradin had gone out of his way to impart fatherly guidance.

At nineteen, I hadn't thought that some fairy godparents could want something beyond good karma and the satisfaction of nurturing a protégé—that even those desires were payoffs. It had never occurred to me that at some level, these were *all* transactions. In life as in economics, indeed.

"COME ON, STACY, let's not end the night," Tristain said in his low voice. We waved good-bye to Wendy and Chito as their cab door shut.

"Okay," I whispered back. I wanted the night to last as long as Tristain would allow. "How about Dupont Circle?" I asked. He nodded, unconcerned, as if the destination were not the point.

To the Metro we went. I felt as though he slung his right arm through my left, but really, we were just close together as we walked across the avenue toward Union Station, the Capitol building majestic and white as moonlight over our shoulders. Down the Metro escalators into the striated cave we went, slipping into the bright glow of the subway car. I suspected he had a girlfriend, but I did not ask. I did not want to know as the chimes rang and the doors closed, each step flowing and happy, each moment elastic, as if capable of slipping us somewhere unexpected.

We leaned back against the orange-cushioned seats, thigh to thigh in the car filled with other kids ready for their night out. The train bustled and rushed, and I liked feeling gently rocked as Tristain talked. He told me what he liked about work and what got on his nerves. I listened, but my attention lasered to the large man directly across from us. He wore hospital scrubs with DC GENERAL stamped on the chest pocket. An orderly's uniform, rumpled, as if he had come from the end of his shift. He did not look at us. He seemed consumed by his task: rubbing his thumb down a pistol clip. Down. Down. His thumb brushing hard against the bullets. I realized that if this man carried a gun to go with that clip, he could rip dotted lines through our torsos. The alcohol I'd consumed slowed time, made me concentrate, made me feel detached. Would something bad happen now? Would our night turn from joy and possibility into violence in just one passage through the city's underground?

"Metro Center," announced the automated female voice. The man stood up and exited our lives.

"Holy shit! Did you see that?" I asked, breathless.

"What are you talking about?" Tristain whispered.

"That guy—the orderly—that was a clip in his hand," I said.

"No way," said Tristain.

"I know! Murder City. I just didn't think on this part of the Metro." I imagined this happening on the Green Line, running down to Anacostia, one of those neighborhoods friends of mine both black and white warned me not to go to. In Detroit, I had run wild but that was my hometown. Here I was new. I stayed in Northwest, thinking the borders would hold—at least hold enough to keep the gunmen at bay.

"Dupont Circle."

Black blanket sky up high. Streetlight haze hovered as the bucket boys drummed in front of Riggs Bank. On the Circle grass sat dreadlocked men next to a few punk-looking girls and boys. Gay men strode by in twos and groups. Searchy kinds of folk passed us too, some rushing, all looking for touch. In government town 1993, Dupont Circle felt like the one bit of wildness where closed worlds opened up, where people openly declared who they wanted and who they wanted to be before they locked it away for work on Monday morning.

"I love it here," said Tristain. So did I. We cut across toward the skinny Starbucks building.

"That's where George lives." I pointed upward, to the apartment above. We saw his red CRX parked on the street. A single light on upstairs.

"Is he still dating Jennifer Grey?" asked Tristain.

"I think so. Maybe they're up there." We both smiled but kept walking. Right then, my life was much more compelling to me than any gossip about my boss. Together we entered my favorite bookstore, Kramerbooks, through the Afterwords restaurant entrance.

I kept moving through the packed aisles until I hit the main entrance table of selected fiction paperbacks, while Tristain remained toward the back of the store. I picked up *The Unbearable Lightness of Being*, hoping he'd see me and ask me about it. The characters stuck with me, and I remembered the scene where Tomas came home to his wife with the smell of the other woman's sex in his hair. Is it possible, I wondered, for lovers to remain both faithful and sexually fulfilled? And if monogamy is possible, is it ideal? I hoped so, for I thought, too, about Milan Kundera's idea of "poetic memory" and how easy it seemed for men to enter mine, to get me to dream about them and desire them as mythic stories, each story ending in union. In such unions, could there be space for more than two?

I wandered and another book caught my eye. It was *Herotica 2*, a book of erotic short stories edited by Susie Bright and Joani

Blank, with a cover illustration of a gorgeous nude woman. I studied the picture. I liked what I saw. This was not the *Playboy* fantasy; this was the best of a real woman, with thighs and belly and B-cup breasts on a warm pink backdrop.

Now Tristain came and looked over my shoulder. "What's this?" he asked. "Erotica," I said, "written by women." Gently, he took the book from my hands. He started reading a story. I couldn't tell which one. He seemed relaxed. As comfortable in his own boy skin as I'd ever known a young man to be. He smiled as he read. I was melting inside, wondering what he was thinking, loving that we were having this moment, whatever it meant. He flipped to another story, and his look changed. "Our society is pretty fucked," he said, "when women fantasize of rape."

"You're right," I said. "I don't think all of the stories are like that, though." I appreciated his empathy, his ability to put himself in our shoes at a time when I was only beginning to do the same for men like him. "But fantasy differs from reality," I said. "In fantasy, you're always in control."

Smiling, I took the book out of his hands, walked to the counter, and bought it. I could feel Tristain watching me. Thinking. I wondered how many scared girls were in his sexual past. Pretty things, but afraid of sexuality. I wanted him to know I wasn't one of those girls. By no means was I expert or even that great a lover yet, but I was open. I wanted to learn. I wanted to please. And though Tristain said nothing more about my purchase, I could feel his mind turning.

CLICK. CLICK. SKIP. Those were my boot heels skidding, skipping down the Church Street sidewalk. *Click. Click. Clop.* Those were Tristain's dress shoes. You could hear every step we made that night. In the city, I felt as if we owned the place. I felt as tall as the town homes with their painted facades.

"Hey, hey, hey, yeah!" . . . I heard only my voice. Loud. Laughing. How does it happen that in the nation's capital, we can walk down a city street before midnight and hear only ourselves? I pushed Tristain out into the road, just playing, of course, not that it mattered, for no taxis raced up this way between those homes and apartments, the ones lived in by the thinkers and creatives of our new city. I giggled when Tristain made to push me back but didn't.

On Seventeenth Street, we passed an aggressive homeless man and a few dart-eyed, wiry boys who looked like gay hustlers. But there on Church Street, we walked in our own noise, snug in our laughter.

Never once did I look up and feel towered over by skyscrapers or tenements. No sidewalk showdowns or competitions for the right of way or a sliver of space to call home. Just peace in my part of Washington, thinking, like so many interns before and after me, that we really could run this town. Easy.

Then you cross the quadrant lines into NE, SE, and Anacostia. Those are the other cities that belong to this city. Their residents cross into the federal district as managers and students and poets and drummers, yet they don't feel like my neighbors; they live in their Washington and I live in mine, and mine happens to be the one that rules the world.

I don't like this, of course. I read that there were schoolchildren in SE who had never seen the White House. Never passed it by. Once. And they were growing up less than five miles away. I was astonished, not yet understanding that this is a common big-city phenomenon, that after the September 11 attacks, we'd know there were Brooklyn and Bronx children who had never seen the World Trade Center, who'd never been to Manhattan at all. How can children grow up in a majestic city and know only their home square blocks? I had seen so much of my city and suburbs as a child. Yet one full year into my Washington experience, I'd never been to SE or Anacostia myself. Carjackings, they said. Gunfights. *Don't go,*

I'd been warned by peers. I listened. As a female alone, I needed to mitigate risks.

But in my heart, I sensed that I wanted to move freely. Connection is what I wanted. Consensual. The kind where no one felt trespassed upon.

THE DRUMS SOUNDED MUFFLED, smaller, the farther up Connecticut Avenue we walked, contemplating whether or not to cut over to Adams Morgan, with its bustling clubs and bars. I wanted to show Tristain a good night out, and back then, that meant multiple spots. But it was late. With my money dwindling, I suggested we go back to Kramerbooks and get tea at the coffee bar. Tristain agreed.

We waited for our order, the music loud throughout the bookstore. I gathered up my courage and asked him the hardest question I could, the one I tried to ignore every day. He looked me in the eyes as we spoke, as if he were one blocked impulse away from reaching for my hand. "Tristain," I asked, "do you have a girlfriend?"

"Yes," he said. "I do."

There it was, that moment the ambiguity dried into dust.

"She's back home in California," he said.

"She's a lucky girl," I said.

He looked into his tea. "I don't know. We've had our problems. But we're trying to work them out."

"I think that's beautiful you're trying to make it work," I said. I meant the words, but they still hurt coming out of my mouth. His relationship with a woman far away whose name I didn't know seemed so unreal. A million miles from the virile young man sitting next to me in a loud bookstore that in five minutes would play The Cure's "Let's Go to Bed" and I would have to leave, insisting that walking home by myself was fine, overwhelmed as I was by a desire I'd been stoking since I was twelve and really began to like boys, certain kinds of boys, to like the type before I ever liked the individual. A connection to the mythic that, at nineteen, suddenly sat only a bar stool away and kept my gaze, letting me touch him that way as we spoke.

THE FOLLOWING SATURDAY, Tristain and I went out with Wendy, just the three of us. We started off at Tiber Creek. We gave Wendy grief about working for a Republican congressman. She told us about the guy culture in her office, how there were lots of jokes that easily slid into sexist, mean riffs and how all she could really do was laugh it off because as a nineteen-year-old intern, there was only so much she could expect to change in her work environment. Wendy would later switch parties and would even become White House staff in the Clinton administration. But back then we teased her, saying she'd made the wrong choice sticking with the GOP. At the White House I rarely felt sexism and had yet to feel racism. Yes, guys flirted, but I never experienced any bullying or anyone being outright mean or hateful to me. I couldn't speak for women higher up the chain, but I thought it meaningful that a young summer intern, in theory one of the most vulnerable people in the system, never had to laugh off a stupid sexist joke cracked by the boss or his friends.

Wendy nodded. To a casual observer, I'm sure she appeared appeasing, but it was clear to me that she would continue on as she wanted. I loved that about Wendy: she had a happy smile but a strong spine. No one was going to push her around.

Talk turned to the president and then to Gennifer Flowers's allegations of a long-term affair, details of which had been printed in the tabloids. "I love that she said he ate pussy like a champ," said Tristain. "That on a scale from one to ten, she gave him a nine. I respect him even more now." Wendy and I laughed. Usually I resisted open sex talk in mixed company, but tonight I liked it. All I know is that with the three of us in a loud bar and him with a girlfriend back home, we could safely push the boundaries.

"Do you think you're a good lover?" I asked.

"Of course," he said, and Wendy and I laughed at his audacity.

"But what about when you were sixteen?" I asked.

"Well . . ." he said.

"Did you have any older women who taught you things?" I asked.

"No," he said, "I was self-taught."

We drank to that.

Drenched in happiness, we left the bar in search of more. A movie? Union Station was just right there. We could—no, let's just walk. Walk till we saw the white glow of the Capitol dome. Truth and justice and the rule of law and all that history. I stared, love singeing hot in my chest.

"Hold on, guys!" I called.

How glorious the Capitol remained all night long, unlike the unlit cathedrals in Europe that become flat-faced mountains in the dark. Inside those halls, this young country was trying to perfect itself, I thought, and we young people were the force to do it. We looked to JFK and RFK and MLK and LBJ and the civil rights legislation and Great Society gains, even 1967's *Loving v. Virginia*, which had struck down miscegenation laws in Virginia. This is what dedicated progressives could do: they could tear down walls that kept us separated, that kept me in this city with more opportunities and you in that city with less—the laws that dared say we couldn't marry and live where we wished! Later in school I would encounter the Jewish idea of *tikkun olam*, a mending of the wrongs of the world one stitch at a time, and I saw our leaders able to do so in those chambers if they chose, with the power to bring about justice one law at a time.

But power, like any force, could hurt. In 1998, I would look at this same Capitol and feel my heart broken, wondering how the men and women inside could pursue these charges against President Clinton for actions that in no way, to me, constituted high crimes and misdemeanors. What a letdown. What a farce. They stood inside their stone structures bathed in light and performed like selfish human beings concerned only with grabbing for more power. But this knowledge would come later. On these summer nights in 1993, we fell in love with the light and white stone, imagining power's possibilities for good.

A few bars later, we called it a night. Walking distance from my house, I guiltily hoped that Wendy would catch a cab, leaving Tristain to walk me home. I had no plan, just hope. Wendy reminded me that she had left things at my place. Damn! I thought, despite knowing better. I asked Tristain if he wanted to come, too. He said sure.

We entered my apartment. I rarely had people over, so I was happy to welcome these two into my efficiency with the dark parquet floor and campaign posters on the walls. Tristain and Wendy sat down cross-legged on the floor and stayed a bit. And I remember something that made me very happy. Tristain asked to look through my photo albums, the ones I had carried from Troy, Michigan, and then from Thurston Hall. He sat with the thick books in his lap and thumbed each cellophaned page slowly. I remember the careful way he looked at all the images and how he sometimes wanted explanations. He could have looked for a few moments, to be polite, then turned the conversation to something else. But he didn't. He took the time to be curious. Tristain may not have wanted me for a girlfriend, but he showed me, by example, that I needed to be patient, that there could be a man out there who wanted to know me and love me as I wanted to know and love him.

They left. An hour later, I received a phone call from Charles. One thirty a.m. Did I want to come over? Just get in a cab—he'd pay for it—and come out to Alexandria. *Just come over. Wouldn't it be nice?*

I told him no. That night, my pride overruled my desire. I'd just had a man like Tristain enjoying my Lollapalooza pictures, and here was Charles calling for sex. I wasn't that lonely.

After that night, I kept things friendly and light with Tristain. The last person I wanted to become was that professor, forcing it.

DURING THE LAST DAYS of his internship, I ran into Tristain on the elevator. He carried a camera. "Hold on," he said. "Let me get a picture, for posterity's sake." I tried to be cool, but my heart leapt. He wanted a picture of me! I might not have been his lover, but I was something to him. Not just another girl. A part of his experience.

Then Tristain walked down one hallway, and I walked down another.

After he left that summer, I never saw him again. But he stayed with me. Just as I hoped I did with the picture, with the thought that maybe I had slipped inside his poetic memory, too.

SUMMER VACATION ENDED. The official interns flew back to their homes, their schools. We had drawn up and disseminated contact lists and promised to stay in touch, if not for friendship, then at least for letters of recommendation. I felt proud of myself: the second session felt like a well-rounded success. We had worked together as a team, and at the end of August, we had old letters still to answer, but we had reduced the backlog so much that Heather was effusive in her gratitude. We had done it. I had done it. My fear that I would let Heather and George down had diminished into a low-level anxiety that I could manage, given that I had already proven myself to be a good office assistant and intern manager in their eyes.

Fall interns arrived as I began my new schedule, one I would keep pretty consistently for the next two years: fifteen credits a semester in my new major, political communication; twenty hours a week as a GWU community service aide, where I was paid $10 an hour to check guests into and out of dorms (and do my homework); and twenty hours a week volunteering for George, leading his intern correspondence effort and helping out in the West Wing whenever Heather or the new staff assistant, Marlene McDonald, and later Emily Lenzner, needed relief.

My junior year, Heather announced her decision to resign. She told us she was ready for new challenges and had set her sights on New York City. Soon she would be doing communications work for Ralph Lauren and would publish a how-to guide for go-getter assistants. Laura Capps would replace Heather. I was not asked but felt no slight, given that I had been asked to become staff before and refused and that I was so deep in my college career that it made little sense to step away before I finished. I'd just earned a spot on an amazing American Jewish Committee study tour of Israel and South Africa. I had been a Truman Scholarship finalist. Since freshman year, I felt as if I had managed to pick the right door to enter, for ever since then I had walked one heckuva charmed conveyor belt. Everything felt so normal and fated and rolling along.

But once Heather left, that ended our contact. She slipped out of the White House biosphere and into her new one, and suddenly the distance between us felt like the light-years between planets. I would later learn that this is a normal thing with professional relationships. When you're not together anymore, you're not together. I missed her and thought of her often, but my plate was full with school and friendships and achievements that reaped unique rewards.

WE STOOD IN A SOHO LOFT, readying ourselves for a rooftop photo shoot. Two long racks of black dresses stood before us, the cocktail kind, with rows of black heels beneath. "Go and find a dress and shoes you like," we were instructed, the "we" being the 1995 cohort of *Glamour*'s Top Ten College Women. I had entered the contest, writing an essay that spoke of my past in Detroit, my present in Washington, and my future in the world as a teller of stories that helped women dream new possibilities for themselves. They chose me that spring, and I joined the other young women aspiring to careers in medicine, business, and law for *Glamour*'s whirlwind winners' weekend in New York City.

The dresses swooshed as we rushed through them, trying to find ones that flattered and fit. This was no easy feat for me, a size 10–12 surrounded by 4s and 6s. I found nothing. I started to feel panicky as girl after girl took her dress into the bathroom to change. Soon just another girl and I remained. A stylist came to the rescue, finding the few options available in our size. Mine had a short chiffon petticoat. I bit my lip, thinking no, this will make me look huge, but I took the offered dress. When they did my updo hairstyle, the stylist smiled—"Ah, Sophia Loren"—and then my worries melted away. But that was later. First, to the bathroom to change. We crammed into the clean space with the natural light from small, high windows. We complimented each other's dress selections. And when I put on mine, a few eyes widened. "What a cool dress," said one girl. I started to relax.

As we zipped and tucked, we talked about who we'd met so far at our *Glamour* events, including three of my heroines: Nobel Prize–winning author Toni Morrison, feminist icon Gloria Steinem, and black feminist philosopher bell hooks. We talked about visiting the Juilliard School, and what it was like to be in the dance studios with the light wood and the barre on the wall, conjuring memories for me of early ballet classes, when I had no idea that I had zero aptitude for classical dance but my mother let me go anyway. "I kept thinking of *Fame*," said one of the girls. And like that, we launched into the anthem for a generation of dreamers, the *Fame* theme song.

All of us singing in the echoey bathroom, including those still checking their makeup in the mirror.

"But wait," I said. "Who remembers that part where Debbie Allen's talking, I've been trying to remember—" Suddenly another girl read my mind, and not only knew what I was talking about, but knew Allen's intro-credits speech word for word: "You've got big dreams. You want fame. Well, fame costs, and right here is where you start paying. In sweat!" Ah! My fellow dreamer remembered! The moment felt even better than the Girls Nation win and its sweeping applause. Two days ago we were strangers, and now here we were in our borrowed dresses, bonding in a bathroom. And on the same page. That simpatico moment gave me the kind of joy that would stick with me forever.

Soon we climbed up onto the roof to take one of those quintessential SoHo shots with the gray sky and building tops as backdrop. This was usually the landscape of action movies, where bad boys ran and leapt, and viewers held their breath to see if they'd fall. But this afternoon, it was us young ladies, pressed together, laughing because now *we* were the It Girls! Most of us laughed ironically, of course, but even if you rejected the fantasy of being put on a pedestal in this way, and had read your feminist writers and knew better than to reach for titles based in beauty and money—powers so easy to lose—who didn't want, for at least one day, to play dress-up and say yes, I exult in all the possibilities available to me? This wasn't the sixties anymore: we had choices. And *Glamour* chose to pick us for our accomplishments, not for our breeding or our fashion sense. Yet we sang in a bathroom and smiled for the camera. Susanne. Rebeccah. Yuri. Cynthia. Azra. Michelle. Nikki. Vanessa. Janis. And me. And when it was all over, I called Ralph. My boyfriend.

Yes—I finally had a boyfriend. After all this time of wanting love, a man finally took me as his and said to the world, *she is mine, and mine alone.* His name was Ralph. I met him in South Beach, Miami, when I was there with my friend Ann Marie for spring break. I was nineteen and he was thirty. "I've got ties older than you," his stepfather said upon meeting me at their Thanksgiving dinner in their Upper West Side co-op. Neither Ralph nor I cared. He remained in Miami, and most of our relationship would be long-distance. But he made me so happy with his *I love you*s and *I need you*s and *you're the most beautiful girl I know*s, which spun me inside until I was dizzy. Ralph had lived so many lives, from teenage boxer to housing loan agent, and he knew how to talk himself in and out of jams. He also knew how to love, for love to me was the nine hours straight he sat at my side in a Miami Beach hospital when I crashed my car during a summer stay. Then Ralph did it again, keeping vigil when I had a terrible flu that the doctors worried was meningitis. He stayed with me through the spinal tap, and through the long hours afterward, waiting for the results. He knew how to comfort me when I needed it, making me feel that he would never leave me while I was vulnerable.

But that summer, Ralph told me he thought he drank too much and that he was going to stop. This came as a shock, because I hadn't detected that his consumption was worse than the partying of any of my college or high school friends. How could the daughter of an alcoholic not know? I hadn't. But Ralph did as he promised: he quit drinking. He praised me for being the force that inspired him to do so. Instead of flattering me, though, his words scared me, for I knew they couldn't be true. I knew that he had done this all on his own. If he believed that I was the key, I worried that this could breed an unhealthy dependence on me. I did not want him to switch one addiction for another.

Ralph also suffered from serious back injuries; one bad pothole on the road could send him into the hospital, racked with pain. He adapted as best he could, as did I, even though knowing he was in pain made me hurt inside. We dated for two years and shared a lot of happiness together. Every accolade, every adventure I had, he wanted to hear about, and I wanted to share, as much as my long-distance calling budget would allow. But as time went on, I became overwhelmed by what I worried could become a huge caregiving load on my part. I never fully admitted this to myself then, but at twenty-one, I was frightened by the specter of my loved one being incapacitated, given what had happened to my own father. Ralph and I broke up, got back together, broke up, got back together. I drifted emotionally. I wanted out of my relationship with this older, charismatic man, so I acted out, developing a crush on my roommate and kissing him, too. Yes, my dream had come true: I had the loving devotion of a boyfriend. But when I had wished for this, I had no idea that the sacrifice and commitment that love called for could loom like thunderheads for a girl with heady plans and ambitions. In 1996, Ralph and I broke up for good.

8

Advice from Vernon Jordan

Washington, D.C.
1995

APPROACH THE BEIGE OFFICE BUILDING, AND YOU MIGHT THINK nothing of it. To the untrained eye, this was just another D.C. mid-rise off Dupont Circle. But look up. See all those floors? That's Akin Gump Strauss Hauer & Feld LLP. That's where you take the fight when you want to win. Or if you need a letter of recommendation from its most famous partner, Vernon Jordan.

I was there to see "Mr. Chairman" in person. After the 1992 election, Jordan served as the chairman of the Clinton Presidential Transition Board, leading the process of creating a new presidential administration after twelve years of GOP rule. Heather affectionately addressed Vernon as "Mr. Chairman" long after that chairmanship was over. He seemed "Chairman of the Board" to me, in the Frank Sinatra sense, for he was so revered, with masculinity that was inspirational for other men and desirable for many women. At sixty, Jordan commanded the rooms he entered, rooms that included the most elite clubs and corporate boards. His titles hinted at but never truly conveyed the power he had, the ears that leaned in to his lips when he spoke.

Yet we could call him Vernon. I understood his allowing this familiarity with George, or George's comrades James Carville, Rahm Emanuel, and Paul Begala, who were all in the tight circle of presidential political advisers. But I was just a twenty-one-year-old intern. To have a man like Vernon Jordan give me permission to call him by his first name felt special.

"Would you like some water or some coffee?" the receptionist asked.

"No, thanks," I said. I looked her in the eye, smiling, for I appreciated the attention. At work, I was usually the caregiver, the one trying to make visitors feel welcome. Rare was a visitor to George's office who didn't wait fifteen minutes in the West Lobby, then a few more in our anteroom. George tried to be on time, but that was the nature of White House life, with meetings running into meetings and high-level staffers juggling simultaneous emergencies, both real and perceived—the difference hard to discern in real time.

Sitting there, I felt the heat rise off my body. The Akin Gump reception area was cool and airy, but I stuck to my clothes. I'd rushed there from the Foggy Bottom Metro on foot in the Washington autumn that still felt like summer. I was wearing flats, thank goodness, so I had arrived on time. Heels made me feel hobbled, as if I should get a handicap score each time I slipped them on, so I avoided them.

But I should have worn my black suit with the sleeker fit. So what if I had worn it to work yesterday? This one was sandstone brown and a little drapy.

I could smell my deodorant.

This was not a modeling audition, yet how often were we cognizant that we were entering a change-making situation? Vernon was as magic as it got, with his power to *poof* a door out of midair, open it for me, and send me on a course that my own preparation, my own class background, could have never conjured on its own. I saw him as a man who, if inspired, could look at me and see something special, in its raw form, and, with his wisdom and his Rolodex, refine it. Launch it out into the world.

For that's why I was there: in a few moments, I would ask Vernon for his recommendation for the Rhodes Scholarship.

Who does she think she is? No one said this to my face, but I still paused inside, each time I thought "Rhodes Scholar" in relation to me and my future, embarrassed each time I mentioned the competition, for it felt tantamount to demanding, "Hey, look at the big brain on me!" No one had whispered those words in my ear as I was growing up. No had one told me to aim for the highest ivory towers. But after my first semester 4.0 GPA and subsequent strong grades, when my political communication professor, Steve Livingston, pulled me aside and said, "Stacy, you should do this," with such conviction in his voice, I thought, well, if he thinks so, maybe I should try. Look what I have to gain.

To be a Rhodes Scholar . . . I coveted the scholar's black robe, for the chance to wear it out in the world, to let it speak for me and my abilities. To wear a robe like a judge's robe. A garment like a shield. If I won this competition, one of the most revered stamps of intellectual approval would be mine for life. *Look at me: If I'm a Rhodes Scholar, how can I be less than good enough?*

But I did not say this out loud. I believed that if you did not draw attention to your weaknesses, others might overlook them. I did not want friends and strangers thinking that I was bothered by the biases of others.

"Miss Parker?"

The secretary led me past the massive dark oil portrait of Vernon that hung larger than life in the hallway outside his office, the one often referenced in magazine profiles. The Big Man. Or the lion in the den. The lion might have been a cliché, but he'd eat you all the same.

There he stood. Sixty years of strength, struggle, gravitas. He looked at me, his black eyes alive.

"Stacy," he said.

The door shut behind me.

FOR TWO YEARS I had interned part-time for George, and every time I entered that small anteroom to take my seat and start answering his phones, I became his first line of defense. *This is Stacy, could you hold, please, thanks . . . This is Stacy, could you hold, please, thanks . . . Hello, this is Stacy.* The phone rang, and it could be anybody on that line, from senior staff to junior staff to journalists to citizens. Regardless, I must be sweet. I must be patient. And since I sat so close to Laura, I must be soft-spoken.

"This is Stacy, could you hold, please?"

"Of course," he said.

"Mr. Jordan!" Even though I could call him Vernon, as George and Laura did, my knee-jerk response was to keep a respectful distance.

"Good evening," he said with his Lou Rawls baritone.

"How are you?" I asked. He proceeded to tell me, and suddenly I heard and saw no one else around me, luxuriating in the full attention he was giving me and I him, a possibility only when he—or really, the president—called or visited. For those moments, everything else fell away.

From the phone display I knew Vernon was calling from the residence. He was very close to the president and First Lady. So was his wife. They vacationed on Martha's Vineyard together, played golf, shared dinners and talks that went into the night. Vernon knew the machinery, the people here in D.C., in a way the first couple from

Arkansas did not. What he gave the president, in terms of guidance, was not quantifiable.

"I would take you out to dinner, but this is tying me up," he said, referring to the current legislative negotiations. He was teasing, I thought, for I doubted he'd take me out that spontaneously. But Vernon had taken me out to meals before. First lunch, then dinner. He had asked me to dinner earlier that summer, and I had said yes, thrilled that this man, who I thought could spend the evening with anyone in D.C. he wanted, would choose to spend it with me. I don't know if he took out any other White House interns or young volunteers, but I had read a mention that he had been seen dining out with a small group of black law students, and I had thought, wow, they were lucky to have Vernon as a mentor.

At a pricey Georgetown restaurant, Vernon made conversation about current events without asking probing questions or making any offers or promises. I remember thinking I wished he had asked me more personal questions, in the way that two friends, when they hit it off, seem hungry for each other's histories. I wanted him to be curious about me. As an uncle might be. Or a father.

But that wasn't the case. We just had dinner. Still, I enjoyed feeling special that this powerful man chose to spend time with me. He hailed me a cab afterward and gave me money to pay for it.

As the cab idled behind us, Vernon said good night with a small kiss on the lips. Quickly—then I was in the cab.

As I was growing up, no relatives talked to me about men. No one pulled me aside and warned me that *men act this way* or *men act that way*. No one told me tales of love turned toxic, of how a smile could be a trick or kisses could come unwanted. My family wasn't like that. I wasn't surrounded by sisters and aunts who had their version of reality they felt they needed to stamp into me. My family would listen and help if I had a problem, for sure. But no one had instructed me on the ways of the other sex, as if it were indeed another species.

Any talk was with friends. In high school, we sailed off eagerly into the world of love, like gentle conquistadors seeking our new worlds. When we were fifteen, this meant hanging around Oakland Mall. My best friend, Peggy, and I were convinced that the salesman at the expensive suit store looked just like the suave, blond, impossibly good-looking Hugo Boss model in the fashion magazines. Now, *that* was the vision of man we most craved, looking exquisite, successful, and virile at the same time. Yes, I had The Smiths in my Walkman and still listened to Nine Inch Nails at home, but I pulled out the color ads of this conservative vision and taped them up on my walls, in my school locker. The Hugo Boss Man.

Only the mean guys who hissed insults at us in our school halls made me worry. Or the loner guys in the mall who followed us from Harmony House to Burger King. Not men like the Hugo Boss Man. Not men who were supposed to have it all.

"YES, GEORGE IS STILL HERE," I told Vernon.

"I'll be right over."

It was late in the day, before the evening news and the latest on the looming showdown between House Speaker Newt Gingrich and President Clinton over the budget. There was a different feel then. Looser, with the headiness of fatigue. Time to order dinner if you were going to stay. Staffers had been at it since 8:00 a.m., some earlier, and I had already had a full day of classes.

But I was no longer languorous. I was excited. I loved it when Vernon visited. He treated me as if I were as interesting as any of the bold-faced staffers around me. Maybe as interesting as POTUS himself.

The president: Vernon would later mention that he told the president we were going out to dinner. I would laugh and say, "Really?" but my thoughts were a cluster of exclamation marks. Flaunting what the powerful can do, I assumed, when not followed by a press pool.

Then I focused on the implicit compliment: the idea that I could be a prize for either of these two men. The idea struck me in a deep place, that same needy place, perhaps, that drove their own incautious actions with women.

I felt someone behind me. I turned around and immediately stood up, for it was Vernon. He hugged me. He made me feel warmly surrounded, small.

"Hello, Stacy," he said.

There was a power to his attention that I couldn't help but enjoy, especially there in the West Wing. People noticed when Vernon noticed you. To some, you counted in a way that you hadn't before. I found those calculations troubling, but that didn't erase their sum total.

Since Laura was at the mess getting George's food and George had stepped out, there were just Vernon and me. I returned to my seat, invited him to sit, but he didn't. Standing there, he had that look that only certain men could achieve with their suits: the perfect fit over the substantial chest, the material that never wrinkled, puckered, or hung.

He talked about the congressional budget fight. He remained standing above me, in that small anteroom meant to hold only one person but carved into two spaces. *Vaseline.* What? *Vaseline, you have to offer them Vaseline. If you're going to screw them, you have to at least offer them* . . . I was smiling, but inside I felt more than exclamation marks. I couldn't believe that he was saying that to me. He was talking to me about Vaseline and screwing.

George arrived. He looked at us, and I could imagine him sizing up the scene: the big man bent over the girl, even though he wasn't bent; he stood tall, as he always did. Vernon gave me a nod before they disappeared behind George's closed door.

Laura returned. I told her they were inside. She hesitated: should she take in the food and possibly disturb them or stay there, with the food growing cold? I shrugged, for I didn't know the answer either.

She knocked, and they asked her to come in. Of course they were nice to her. They were gentlemen above all else, and gentlemen were never upset when a beautiful blonde walked through the door.

The phone rang, and I was talking, but I wasn't listening. I was thinking about Vernon. I knew he'd taken a liberty with me. People didn't usually talk that way in our office. If language grew rude, it was behind closed doors. Never in my presence. Never directed to me.

I thought about the scene, and his words bothered me. They conveyed a certain crudeness and aggression that he'd never directed at me before, different from the good-bye kiss beside the cab. I could still read that cab moment as an innocent display of affection. Nothing in that anteroom moment had felt affectionate. It felt mean and cold.

I knew, however, that if I were going to get anywhere in this town I had to ignore this kind of thing—at least ignore anything that was bearable, and a Vaseline remark was bearable. One bad moment, I felt, should never shut down a relationship.

But I felt the shift inside me, the experience as a warning: this man was no fantasy father, and I needed to take heed.

VERNON HUGGED ME upon my entrance to his Akin Gump office. He pulled back and with total confidence went for a light kiss on the lips. I moved my face, but I was not fast enough. I would have preferred not to greet that way.

He motioned for me to sit. I did. I saw the coffee-table book, the one that featured portraits of fathers and daughters. I knew he was in that book with his daughter. He sat next to me on his couch. I noticed that he gave us space, that he did not crowd me.

But he glanced at my legs.

"Stacy," he said, looking directly at me. "Always wear hose."

"But it's still summer . . ."

The words came out of my mouth before I could pull them back. Perhaps he had no business giving me grooming tips, but protesting advice was bad form regardless, and using the weather as an excuse not to conform to etiquette was just stupid, because I knew that in his mind, it didn't matter. For a young black woman with aspirations, there could be no chink in the armor. I knew this was the same thinking that had created Motown's "Artistic Development" classes in manners and elocution. That if the talent came out of Detroit housing projects or from other déclassé origins, Berry Gordy wanted no embarrassments, no gaffes that might be used as "proof" of inferiority. He wanted high-gloss finish. His artists would know the rules of the game, the etiquette that proved you belonged at the same tables as the best of the whites. Jordan and Gordy both seemed to believe that clothes and etiquette were the strongest armor, the best-sealed shell to slip our full selves inside. I happened to like most of my full self and felt accepted by the people who loved me, flaws and all. But I knew many folks, no matter their background, never felt that comfort.

In the meantime, I could reassure Vernon Jordan that I valued his advice, even if I didn't fully agree.

"I will," I said.

Then Vernon asked what he could do for me. I talked, but I was outside my body, too. This big man in the elegant suit with the deepest brown skin had made time in his schedule to sit with me, to allow me to ask for help. This man with the eyes that could hold you and meat-hook you, too.

I asked him for the recommendation.

He agreed easily. I should send him a draft and he'd take it from there. This was not my favorite method, for it was odd writing out good words you wished would come naturally from the man's mouth, but it was to be expected. A man like Vernon (or his secretary) didn't have the time or desire to draft a letter like this from scratch. It was the yes that counted.

Our time was up. He hugged me one more time. This was perhaps more dangerous, because we were alone in his office with the door firmly closed. I never expected him to be aggressive. But one wrong look on my face, one millisecond that everything was going too far and we both knew it and the fear that instead of apologizing, he would blame me somehow, that I would become the source of shame—the risk felt real. He might have had nothing but honorable intentions, but it was hard to know at that moment. I was young, a novice at gauging just how much strength a man would use to get his way.

He dry kissed me again.

I pulled away gently, and he didn't resist.

I thanked him for his time, his help. He opened the door and slowly walked me to the elevator. There was nothing about him that said his mind was on the next phone call or the next meeting. He still felt present. He said he was glad to do this for me. I saw no doubt in his eyes.

But five minutes down Connecticut Avenue, my stomach felt unsettled. Not sick, but anxious somehow. As I felt after the Vaseline quips. As I felt when I'd been before a man who lusted for me but did not love me.

What was it with those kisses! In Washington, D.C., I found that many men did this—older men of stature, but also younger men trying to emulate them. Once a White House staffer only ten years older than me had tried to greet me this way, a man as pale as a vampire, and I had thought, I can't believe it, you too? You thought you were giving them your cheek, and *pow!*, there they were, right on your mouth. They acted as if it were okay, as if we were all such close friends.

You didn't wish to offend. You wanted to feel as if you belonged. You needed them to help you, to protect you, not to hate you. You pretended it was okay. And sometimes it was okay, especially if the older men were kind.

When Vernon saw me again, he acted gallant. Later, I would accept another invitation for dinner that would again be uneventful. I imagined that our meeting in his office had made no more than a blip in his consciousness.

Not true for me.

That day, I hustled down Connecticut Avenue, walking toward the White House for my afternoon shift. If I kept at that speed, I'd be only ten minutes late by time I got through the northwest gate and walked up the drive past the journalists and their stand-up positions. I checked my reflection in a restaurant's plate glass window. I didn't look polished, but I looked okay. My hair was frizzy at the ends, but I'd live.

No, I didn't duck into the Dupont Circle CVS to buy cheap emergency hose. It was simply too hot.

But Vernon's words stayed on my mind. Beyond the recommendation and the dry kisses, he had dangled that potent fantasy before me. First the hose; then soon, Stacy Parker could *finally* be a certain kind of girl, one of the perfectly groomed. Me, only polished. Like the horses I used to ride as a child back in Michigan. The round brushing. The cleaning and the caring. All of that tender corporal upkeep. Listen to the change maker, and suddenly I too would command rooms. I would wield the power of the elites.

But first, becoming the perfect young lady required learning the steps. The stuff of the ring: of the whip in the hand and the heel in the side.

My grandmother had tried to instill rules in me, long ago, before we had moved out of the city. Nana had bought the scratchy dresses I rarely wore. "Put your napkin on your lap," she would say.

"Don't slurp." "Only girls with clean fingernails can wear polish." There were other instructions, too. I could feel them, pushed deep into my subconscious. *Do this*, if you want praise. *Do that*, and you would bring blame on yourself, or the worst kind of shame.

I had a memory that may seem like nothing but to me was very telling. I was five, in the back of Nana's black Cougar, falling asleep. She turned back and asked, in a baby-talk voice: "Are you sleepy?"

"No," I said as something flared inside me. "I'm *tired*."

I bristled at the baby talk, as if it were a silken muzzle that could wrap around my face, my voice, and my will. Keep me someone's baby for good when all I wanted to do was grow into the girl, into the woman that was uniquely me. A process that required experimentation on my part. Trial and error and not always following my elders' instructions.

Yet I knew she wanted the best for me. Her wisdom was hard-won; she knew exactly what it was like for a young black woman to make her way in this world, the biases she must clear like fences, every day, every hour, just to get into the house through the front door. Though my grandmother came from a time when elders didn't necessarily explain their edicts, and she never explained hers to me, I never doubted she had learned from experience.

But I already felt, I knew, I hoped that there could be another way. That I was me, Stacy Parker. I didn't want to be a clone of someone else. That although winning the Rhodes Scholarship would be an exceptional honor, I could also navigate my way without it. I could be good enough on my own. For two years, I hadn't even been an official White House intern. That had showed me how far one could go without a title.

The uniformed guards smiled at me through the northwest gate window. I fumbled for my hard pass. Once again, I was digging through my bag, feeling the crud catch beneath my fingernails. There was no shade on Pennsylvania Avenue. My cheeks got pinker in the heat. I looked alive, happier that way. I knew that if I had been wearing hose, I would have been miserable. I knew that if I had been wearing heels, I wouldn't have made it at all.

VERNON MAY HAVE BEEN the most famous man to give me guidance my senior year, but he was not the only impressive person to do so. From Dr. Geri Rypkema, the director of the Office of Graduate Fellowships, who shepherded me through the Rhodes process, to Dr. Jarol Manheim, the founding director of the School of Media and Public Affairs, who supervised my senior thesis on media coverage of the Bosnian conflict, to Pulitzer Prize–winning journalist Haynes Johnson, who knew how to put all of this work into perspective, I felt surrounded by accomplished educators who seemed invested in my success.

But the GWU professor who had the most sustained influence on me was Dr. Steven Livingston. He taught my sophomore Intro to Political Communication class, as well as later courses, and continued to mentor me long past graduation. I will never forget him telling me what a strong writer I was, one of the best he had graded up to that point. Those words built my confidence and helped me know with certainty that I could compete in the academic world, if I so chose.

Originally from Flint, Michigan, the depressed auto town less than an hour north of Troy, Dr. Livingston, like both of my parents, had left his birthplace early via the U.S. Army. Dr. Livingston later spent several years in Washington State and grew to think of that rainy green country as his spiritual home. Still, I felt our shared rust-belt origins connected us. We both needed to leave home to "be all that we could be."

Dr. Livingston was handsome, with gold-rimmed glasses and chestnut brown hair. In his early forties, he brimmed with energy: every semester he seemed to be publishing something new in our expanding field of media studies. I remember walking into his office and waiting until he finished the last line of a letter—maybe my very own recommendation letter. Once he finished, I had his full attention.

"How are you?" he asked.

Nostalgic. I was almost done with undergraduate work. Soon I'd no longer hear his voice from the lectern, in the small darkened hall where we looked down on him while his words filled our ears, back when he reminded me of a broadcaster, given how full his voice could sound as it carried his lecture so dense with knowledge. Maybe he sensed the end coming too, for he asked, "Are you in a rush? I know this is going to seem out of the blue, but how about we take a walk? Have you seen the cherry blossoms?"

Was he serious? At twenty-one, I had never seen the cherry blossoms, nor the crowds of tourists and locals that slowly circled the tidal basin in front of the Jefferson Memorial, snapping their pictures. School and work had seemed so urgent. I'd blown off this beauty, the chance to linger beneath the clusters of pink-white silk.

My professor for Asian Humanities, Dr. Jonathan Chaves, had introduced us to ancient Japanese poets who saw their human experience reflected in the nature around them. *Consider the plum blossom on the branch. There is so much to be felt!* When he talked beauty with us, he was not melodramatic, but he seemed to experience altered consciousness when considering those things that gave him, and the ancient poets, pleasure. You could almost hear a gentle chord change in his voice. He shared with us the Japanese concept of *mono no aware,* "a sensitivity to things," the ability to behold a blossom bursting forth or in decay and have that transfer of understanding, as if its meaning were mainlined to one's heart or mind without need for explanation. *Consider the dried petals fallen to the ground.* Past Japanese aesthetes declared the ability to reap meaning this way the mark of Japanese singularity and superiority. Although I was not Japanese, I came from a city of ruins, and I had seen my interior life etched in the painted bricks turned colors by sun, rain, and neglect, and in the houses warm with light. No, I wasn't Japanese, but *mono no aware* stayed with me. "Okay," I said. "Let's go."

Dr. Livingston and I visited on a weekday, so the experience was not marred by weekend crowds moving shoulder to shoulder. We could breathe. We walked slowly, taking in all of the beauty around us. Dr. Livingston had spent several class hours teaching us about the workings of power. But on that day, he showed me the power of enjoying what good was around us, waiting for us to simply stop and experience it. And in those blossoms so abundant, I saw that my life was beautiful, and that despite the occasional strange moments with mentors, my path was protected. I felt nurtured and cared for, in loco parentis.

I DID NOT WIN the Rhodes Scholarship. However, I did progress to the regional district finals, and afterward my school awarded me its Shapiro Scholarship to Oxford. In the fall of 1996, I would travel to England and read politics as a visiting student at Hertford College. But that was months away. In the meantime, there was the president's reelection campaign. As I finished my third year of service in George's office, I thought it was time to try something new. I spoke to George, and he placed a call to the White House advance office, and soon I would begin what appeared to be the most glamorous kind of campaign work: presidential advance. Traveling from city to city, state to state, advance people did the prep work, then ran the trips, whenever the president or First Lady traveled. Every young staffer and intern I knew wanted to do advance work. I couldn't wait to get on the road.

9

The Trouble with Agents

I RAN INTO MY HOTEL ROOM, HAVING HEARD THE PHONE RING
from the hallway. Yes, it was past midnight, but everyone knew
time didn't matter much on presidential trips. Staffers worked late
into the night and rose early in the morning to ensure that the trip
ran smoothly. In July 1996, I had joined the pool of advance people
who traveled the country for the president's reelection campaign.
Advance people ran sites, press operations, the motorcade, and the
hotel if the president remained overnight. No matter our advance
staff titles, our duty was to be available as needed. *Get a thousand
more people into this arena. Drive these staffers to the good gym across town.
Figure out how to order fifty TVs.* Tasks were tasks, and I was there to
complete them, still feeling as if each one could make or break my
future in that world. Before, Heather and George had known my
work pretty well, and a mistake would be just that: a mistake. But
on a presidential trip, I was assisting all the rest of the senior staff.

All they would know of me was how well I performed their assigned chores.

Three rings, and I made it to the dresser and pulled the receiver to my face.

"Hello, this is Stacy," I said.

"I saw you downstairs," a male voice replied.

"Who is this?"

"Such a pretty dress you were wearing," he continued.

"Is this Brian?"*

"Why don't you come upstairs?"

"I can't," I said.

"C'mon. Don't you want to see me?"

"I have to be up in three hours. I've got clips to do."

"Come upstairs."

"I'm sorry—"

"Please . . ."

"I don't even know you," I said. "Good night, Brian."

Whatever he said was muffled as I cut off the call. The AC was running too high for the night's coolness, and I sat there, chilled, staring at the beige plastic phone. A man I barely knew thought it was okay to call me after midnight. A married man. A Secret Service man entrusted to protect but trained to harm if necessary. *What in the world was he thinking?*

Ten minutes earlier, I had left the staff office—a large, converted hotel sleeping room—and realized I had left my key on my TV and I needed to go downstairs to get a new one. In our atrium hotel you could see every door from every floor. Brian must have been watching from upstairs. Watching from his floor above mine.

I imagined he had a master key. Stephanie* did; she was the RON I to my RON II, both of us tasked to oversee hotel logistics. Brian was our Secret Service counterpart. If Stephanie had a key, Brian did, too.

---

*   This name has been changed.

Not that Brian needed one. The midnight desk people would give any one of us keys since our party "owned" the hotel for this trip, especially if the requester looked like Brian: clean-cut and handsome, if in a plain way. What you'd want in a young government officer. Yes, if a man like Brian went down and asked for a duplicate key, no matter what the room, the desk staffer wouldn't think twice. She'd just do it. She'd reach down, slip the plastic into the machine, and give it to him.

I got up and turned the bolt.

THIS WAS MY LAST ADVANCE TRIP before leaving for my fellowship at Oxford. The advance office took a while to slate me, but once it did, I was thrown in; my first trip was as the RON in Little Rock after the postconvention bus trip. Twenty-four-hours-a-day doom, that's how that trip felt, as I didn't have much guidance beyond the advance handbook, the phone numbers of two other RONs, and a few snatched moments of instruction from my lead. The president overnighted at a private residence, so I did not work with him. But the staff arrived at the sold-out Excelsior Hotel, and I discovered that the speechwriter Michael Waldman was with them but not on my latest manifest, which meant he had no room. After much scrambling, the manager gave up his room to Waldman. I thought the advance office would never slate me again. But when I saw my lead, David Neslen, after wheels up and I told him this fear, he laughed and said not to worry.

Neslen was right. I was slated again, this time as the RON II. As RON I, Stephanie dealt with the Secret Service. She attended the walk-throughs, negotiating whatever was left unsettled by the lead advance. As RON II, I got stuck with much of the schlep work, including clips duty, which had me up in the middle of the night making sure that the morning's news clips faxed in from the White House arrived and were photocopied and distributed to the president and key staff. I survived on three hours of sleep per night. I lost weight, dropping more than ten pounds. I was pushing a size 8. Slim in a way I hadn't been since high school. I think I looked okay before, but I'm sure that to some guys I looked better.

Brian performed the RON duties for the Secret Service, overseeing security for the hotel. I had no professional interaction with him. We had met for the first time only two days before, sitting outside one of the president's events inside the hotel.

Out of friendliness—and boredom—I started talking to Brian. About nothing. Chitchat, really. But as time dragged on, I started asking Brian more personal questions.

"Are you a father?" I asked.

"Yes," he said.

"Oh! Do you have pictures?"

"Yes, I do."

He made a slow reach for his wallet, then showed me a picture of his wife and two children.

"You have a beautiful family," I said.

Brian nodded, smiled, put the wallet back into his pocket.

"So, may I ask how old you are?"

"Thirty-nine."

"No way! I would have *never* guessed you were that age."

He looked surprised. As if he could not believe my surprise.

"I mean, you just look so much younger!"

He turned toward me. I noticed him look me in the eyes for the first time.

"What's not to believe?" he asked.

"It's just . . . I mean . . . you're handsome in a youthful way, I guess."

"You're just saying that," he said.

"No! No!" I said, but I wanted to start over. I felt as if I were out on a long, thin limb. Not that my words weren't true, but I had no interest in leading on a married man.

The door of the conference room opened. The president would move to the next destination on the schedule. Back to work we went. I don't remember any more words between us.

Until he called me in my room, inviting me upstairs.

Even if he had been single, I would still have hesitated. Older staff women had warned me about agents. They're like ballplayers on the road, they said. Fuck one, and you're a notch on a bedpost. Love one, and you cry.

FOR AN AGENT, standing post can be the worst. Hours alone in a stairwell. In a cornfield. In a dark place that gets darker than you've ever known if you're from the city. The principal goes on vacation, but he picks the bush, not the beach, and there you are in the Texas backwoods. Nighttime comes, and they're cozy in the ranch but you're standing outside in an open field with a million bitty creatures buzzing and stuttering their codes of desire, trying to break your skin.

But you're quiet. Aching. Standing post is hard on your feet, on your back, too. Think of all that stuff you're carrying. Your radio. Your weapon. You feel it when you're standing for hours at a time, and it doesn't matter how fit you are: stand there long enough with all that hanging from you, and your body talks. So does your mind.

Once you were on the CAT team, trained in counterassault. Back then you had a partner and you could talk. About nothing, sure, but talk kept you even. Now you're on the shift. Now you're out here by yourself, wishing you'd bought some Kodiak dip at the gas station as your mind runs lap after lap.

You vowed you'd stay true. You vowed you wouldn't be one of the guys in divorce court. You wanted to go on the Presidential Protective Division. Get it over with, you told your wife. Get through it and move up the ladder. She wanted that bigger house, right? This is the job and what they expect us to do. Just hang in there. We'll get through this. Together.

But right now you are alone in Texas country with no moon above, the land at your feet more alive than the wildest city block. You're not so removed from your country roots. One generation? Two? Yet this is the worst kind of amnesia. You can't just summon up that wisdom by ancestral osmosis. Maybe your father taught you how to survive. But you were city boys. He didn't teach you what to do when you're stuck for eight hours in blackness as abandoned as anything you've ever known. "Welcome to the jungle," you sing to yourself. You try to laugh it off.

You've heard the stories. Stand post out here through the dark of a summer night and you might lose it. You might just pull your gun. At the command post, they will find out and laugh. Especially the southern agents, the ones who've hunted and fished their whole lives, the ones who know woods, deep woods, and every line of "A Country Boy Can Survive." They're the ones who default to Hank Williams, Jr., while you're stuck with Axl Rose.

If only the girl would look at you when you return to the hotel. Everything would be fine if she looked at you. Think of the girl, and you feel better. Think of the girl, and you make it through your shift.

THE FOLLOWING NIGHT, I slipped the next day's final schedule beneath Brian's door. He opened the door seconds after the paper left my hand. As if he'd known I was coming.

"Hello," he said.

He stood there in a white T-shirt and gym shorts. He had tanned, athletic legs downy with hair.

"Come in," he said.

"No, I have to finish this, then get some sleep."

"Are you sure?" he asked. "We could have a drink."

"No, thanks," I said.

"Are you sure?" he asked.

"I have to go," I said.

I kept my resistance curt, refusing to play coy or sweet so he would not think my "no" meant yes. Brian was a good-looking man, but I had to go. Once again it was after midnight, and I had no business in his room. Worse, the whole hotel could see us. A hundred doors could open and watch the whole scene.

"I've been waiting all night for you," he said.

"Sorry," I said. And that's when I glimpsed his gun on the nightstand, matte onyx under the lamplight. Would he use it? Probably not, but I didn't know. The gun didn't even matter; if he was that kind of man, his trained hands would do.

I continued down the hall. I felt his eyes watch my backside. Then the door closed. I turned around and saw nothing but the unnatural light of an atrium lobby lit up at night.

Like a hothouse. Yes, this was a hothouse: how else to explain the flirtations, the come-ons, the sex that had to be happening with all these driven people pressed together for days, each room furnished with double and king-sized beds? As if you could leave consequences behind you as you would a wet towel on the bathroom floor.

I had already hooked up with one advance-team staffer. Just one night, after we had been drinking. Then the road show had arrived. Twice I had found myself delivering papers and invited into the rooms of traveling staffers. Quickly, of course. When I stood in the doorway, everyone could see. *How are you? Come inside.* I walked in, knowing it was strange and dangerous but flattered to be paid attention to by men I respected. Twice I found myself invited to sit on the small couches of their junior suites. They talked to me sweetly. They made their intentions felt.

Twice I got myself out of there. Gently. Quickly. Out. Both men were married. Both men were laughing as if it were funny.

One leaned forward to kiss me, but I pulled away. Just kidding, he said.

Just kidding, my ass.

I WENT OUT with an agent once. Joe, the Puerto Rican uniformed guard who sat at the Seventeenth and G Street entrance the summer I first started interning at the White House. He always bantered with me sweetly, energetically, welcoming me to his desk with a ready smile and laugh. I felt as if we were friends. When he asked if I wanted to see the new Secret Service movie, *In the Line of Fire*, I said yes. Sure, he was older and an agent, but this could be whatever we wanted: two friends seeing a movie or two adults on a date. There was safety in ambiguity.

A few nights later, Joe and I walked into the movie theater a few blocks from my Virginia Avenue apartment. Sitting down, I liked being close to his body; the heat of his forearm pressed against mine on the armrest, and I loved how that could be construed to mean so much or nothing at all. Clint Eastwood stood above us, strong and troubled, but in the velveteen seats, I felt excited, even worried, especially during the scene where they rushed the president out through the kitchen. I wanted the president to be safe and I wanted the agents to succeed, even if this was just a movie.

Afterward, I listened to Joe assess the film's realism—approving of some moments and dismissing others—as we drove back to my apartment. Joe parked in a visitor parking spot. He jumped out quickly so as to open my door for me. His fingers held my own as he helped me out.

When he hugged me, his body felt so good against mine. I wanted a kiss. "It will drive me crazy," I said, "if we don't"—and he kissed me. Hands holding my shoulders, then fingers to my face.

But that was it. He didn't come upstairs, and we didn't go out again. I think Joe had a girlfriend—or several, who knew? I was still trying to figure out the Charles situation. Honestly, I don't remember why nothing more happened between Joe and me. I was probably too young for him. Regardless, he never took advantage, and he never strung me along. With Joe, I never felt used or thrown away.

WHEELS UP. The president and entourage flew off to the next campaign stop, and with that, I felt the deepest relief. Stephanie might have been RON I, but any failure would have been my own. No one had ever run me this hard. I was ready to drop.

Only a skeleton crew of us remained to pack up the offices and settle the bills. After all the push push push to prepare and run the trip, the last thing I felt like doing was cleaning up messes and getting together names for official thank-you letters. I felt as if we'd crossed the finish line but still had more running to do.

I had to hurry. I had a plane of my own to catch. If I missed this flight out, I would miss my London flight from Detroit and screw up my Oxford arrival.

I ran up the hotel stairs, chutes of gray concrete and white-painted walls reeking of bleach. I rounded a landing and—Brian.

He stopped. So did I.

"Hello," he said.

"Hello," I said back, my heart beating hard from the ascension. And from him.

"I was hoping I'd get to see you."

"I have to finish—"

Before I could say "tearing down the office," he walked down the two steps to the landing. Little separated us. I was wearing that dress again, the one he had noticed that night. After his late-night call, I had tried to be short with him, but professional, doing what I could to telegraph that nothing sexual was happening between us.

His gaze lowered as he looked at my breasts. I wanted to slap him, to tell him to keep his eyes where they belonged and get out of the way, but with a quick pounce his body pushed mine against the wall. He leaned in hard, his belt and badge pressed into my waist, his thigh prying open my legs. With one motion he gripped my wrists and pinned them above my head. His mouth was pushing, trying to get my mouth open. I refused. My body was one big NO while my brain struggled to catch up. I felt the free fall of the moment. I thought *wife, children.*

He pushed everything against me, his mouth hard, everything hard.

I opened my mouth.

His grip softened around my wrists as we kissed. Then I closed my mouth again and pulled out from under him. He let me go.

"I've been wanting to do that since I first saw you," he said. "I know you've wanted me to do that too."

I did not look at him. Instead I walked straight past him.

"I have to go back to work," I said.

"I'll be in my room if you want to see me."

"Okay."

I didn't want to say "okay," but I wanted to keep my exit greased and moving.

I pushed open the next door. Not the floor I needed, but that did not matter. I took the elevator the rest of the way.

Hours later, and I still wanted to know *what the fuck was that?* I shook inside. My mind raked over every moment. That was the first time I had ever been attacked.

Attacked is a strong word. He had let me leave. He had only kissed me. And yes, in the end, I did not hate it. If he had been single, if I had not seen the picture of his family, maybe I would have stayed pinned against that wall. Yet the fact remained that he had forced himself on me. He was an agent. I was entry-level campaign staff. He wore a gun. I did not. He had brute strength. I was too young to gauge how far he would go.

I had never given him a green light of any kind, I thought.

Was this all because I told him he looked young? In my naiveté, had I shown too much personal interest in a stranger?

That had to have been it. He must have thought that was the opening.

Unbelievable, I thought. What kind of world was this where a young woman couldn't be friendly to a man without his thinking she wanted him sexually? If that was how the world really was, I didn't know what I was going to do.

I finished my work, my packing, my trip. I boarded the shuttle bus for the airport. I missed the flight, but with some quickstepping in Detroit, with my grandmother getting me home to get the rest of my luggage and back to the airport, I was able to make my flight to England. I would not see Brian again for two years. And when I did, it was only in passing in an OEOB hallway. I was not even sure at first that it was him. We did not say anything to each other. Not then. Not ever again.

WHEN I HOOKED UP with the advance guy that night on the road, he pulled me toward his hotel room. Quickly, for that was an atrium hotel and everyone could watch. I did not resist. He was a boy in his twenties. So many young staffers want to be so adult, but they come off like kids. One staff guy called another staff guy "lint" because he always had dust on his suit. Men like boys who wanted to be masters of this universe but who knew how to bitch behind one another's backs. Boys who weren't like Rahm or George, who were men in my eyes and unavailable in their adult relationships. Boys who were still boys.

I judged my advance peers harshly, but what exactly was I? A girl! But a powerful one, only peripherally aware that I wore the body of a woman and that this body had its own power to rival any other force on this earth—that is, if you were a nearby man and you were hungry and lonely. I might have pointed to bits of dimpled fat in my thighs and thought, geez, how powerful a sex object could I be? But the men saw the whole, or they saw the parts they liked: the breasts, the eyes, the hips right for holding. This wasn't just me. All of us young women were powerful in ways most of us could not imagine, unless we thought about the magic we attributed to men, how we insisted on believing that the right boyfriend or husband would radically transform our lives, even though they were only men. Only people like us.

The trouble with girls is that we don't always know our power. We know men have the upper hand so often that we assume our power is not potent. So some of us play and push. Play and push. Some of us coy, some of us oblivious.

So take a girl like me, and surround her with agents. The Erics, the Brians, the Steves of the world. The Darryls and Clarences, too. Gorgeous. Polished and trained. Each day I walked the West Wing, I saw these alert men standing, sitting, greeting me as I passed their posts. On each trip I worked with men who had to maximize all of their gifts, physical and intellectual, thrown into so many different situations as they were. Some days guarding a door. Other days managing the logistics of an entire leg of a trip abroad. I saw them in the hotel gyms. They ran on the treadmills while I worked the StairMaster. They came around, encouraging me, pleased that I took my physical fitness seriously. Later, I would learn that when you partnered with them at meetings with hotel management, or with overseas police chiefs, you could play good cop/bad cop, you the smart, well-spoken woman and he the silent, strong man who lets you make the case, then backs you up if the resistance is too tough; who, if need be, goes eye to eye, tone to tone, two strong men sizing each other up. And since your strong man is U.S. Secret Service, guess who's going to win?

Imagine going upstairs with him, afterward, to his room. Off with the blazer. Off with the radio and the wires. Put the phone, the pager, every bit of electronica to the side. Off with his piece. Now he's just a man. Hard-chested, hard-thighed, and standing there. Telling you to come here.

That's the trouble with agents. Once you get that far, you're through.

All I knew is that I kept looking to these men for something. And when moments got scary, I did as Mr. Bradin had advised: I slipped out of the scenes before they spiraled out of control.

Good advice for any girl traveling the country as an advance person. Or traveling abroad as an international student, which I was now at Hertford College, University of Oxford.

CAT ON A HOT TIN ROOF. Maggie the Cat in her silk slip, spurned by her alcoholic, heartbroken husband, Brick, and slinking up next to Big Daddy, still so formidable despite the cancer, the patriarch who says that if he were Brick he would have had Maggie pregnant long ago—look how Tennessee Williams laid bare these human dynamics in the skeletal frame of a play's dialogue and directions! I sat in the dark wooden carrel in the Hertford College basement library, my philosophy of law book uncracked. I had spied the small paperback in the stacks, and I read through the whole play in one sitting. Then *A Streetcar Named Desire.* Then *Suddenly Last Summer.* I'd known of Tennessee Williams, but I'd never read or seen any of his work. Williams showed me what happened when a woman like Cat wanted love and her husband, Brick, wanted his dead best friend. He revealed the family as machinery: push that button and watch that smile, pull that lever and hear that scream. I thought this was brilliant, so vital, the way to the mutual understanding that my progressive friends and I thought laws could deliver. If you see why people do what they do, you can sympathize. You can soften your heart. As Shakespeare wrote in *Hamlet,* the play's the thing to reach the king—and villagers alike. My philosophy of law book remained closed as I felt this knowledge sink in deeper: I'd much rather be a good storyteller than a good lawmaker.

Maybe I could tell my story of agents, of me and these men and these pressure-cooker situations, and readers would understand our failings better. I imagined that if each of us spent more time entering others' minds and motivations, fewer mistakes would be made in our own lives, too.

In the meantime, I read politics and international affairs in Oxford, England, the ancient university town surrounded by green countryside that in late spring would seem drunk on its own lushness. I settled into a group house on Leckford Road, the same road that President Clinton lived on when he was a Roadie, and spent my first weeks reading and writing and using the computers in the Octagon with other graduate students, including two Americans, Stephen DeBerry and Adam Smith. We disrupted our work to debate aspects of Swedish-style socialism and whether we wanted to live in the philosopher Robert Nozick's "experience machine." Together we kept up with the political world from our lofty new perch. I wrote essays and read them, per the tutorial style, to my don, Nigel, who himself had recently earned his PhD in political science at Harvard University. He seemed to love working with an enthusiastic learner who had come straight from the campaign trail, and I loved working with a thoughtful scholar who would grow to be a dear friend.

I had spent my work and school life studying the ways of power. I had tried to understand the older staffers and mentors around me and the thin lines between nurture and control and manipulation. Now that I had lived in England and traveled to South Africa, I was deeply curious about systems of force used in other countries. I decided to apply for a master of studies in modern history, focusing on British Commonwealth history from 1860 to 1960. Following the lead of the power scholars Stephen Lukes and John Gaventa, I wished to examine how quiescence had been maintained and dominion challenged within the Commonwealth, with special attention paid to South Africa.

I was turned down. Perhaps my judges looked at my transcript and thought I didn't have a deep enough background in the subject matter, given my transcript loaded with communications, humanities, and writing workshops, courses that could be dismissed as lightweight by some academics. Perhaps they thought I didn't have the academic chops, period. Regardless, my school career came to a halt. After a fantastic year of study, of making new friends, of even falling in love with a young, beautiful English student, Jonathan, I had to make a choice: to stay with my love somehow, even though I had no school placement or work visa, or to go home and begin the professional life for which all this schooling and White House work had served as preparation.

I had no work visa. I had no money. I loved Jonathan, but the choice felt clear: I had to go home. In June 1997, I returned to Washington, D.C.

II

STAFF

10

Welcome Back, Swimmer

Washington, D.C.
1997

THIS USED TO BE PARADISE CITY FOR A YOUNG WOMAN LIKE ME brought in to help, unpaid. The distinguished journalist Al Hunt had once chatted with me in George's anteroom as he waited to meet George. He'd looked starry-eyed, contemplating aloud how wonderful it must be to be an intern just feet away from the Oval Office. What a future I must have in store. White House interns rode the fast track to truly being somebody. He'd talked and I'd smiled, believing all he said to be true.

Two years later I stood at the northwest gate, waiting as the uniformed guard searched for my name in the Worker and Visitor Entrance System (WAVES). Upon departure the previous year, I had surrendered my pass, so I was like all the others who needed confirmation that they had permission to enter. My old pass had given me instant access. I would hold it up to the sensor and walk past the turnstile as easily as any agent or the chief of staff himself.

Not today. At least not yet, for this was my first day back at the White House. I returned, no longer an intern but staff. I had finally breached that barrier, slipping into the rarefied world I had served hungrily for three years prior. A world that wasn't the same as the one I had left. Not only had George resigned after the 1996 campaign, he was now a broadcast journalist for ABC News, appearing on the Sunday-morning news show. My new boss, Paul Begala, knew me, but the staff people who had known me best, Heather and George, were both long gone. Even my friend Justin had moved with Mike to Seattle.

As I stood there in the morning sun, my mind resumed its pinball game, the ball bouncing, then ricocheting against so many worries. Now I was no longer a tabula rasa. Fear grew when you knew the myriad ways to look incompetent, to let your boss down, to look as if you didn't belong.

The street behind me felt parklike, with the birds calling and flying overhead. Pennsylvania Avenue used to have traffic: taxis and tourists and everyday people driving past. Not anymore. Not since Oklahoma City in 1995, when Timothy McVeigh had blown up a Ryder truck full of explosives next to the Alfred P. Murrah Federal Building, taking 168 lives, including those of 19 children. The president had spoken at the memorial prayer service: "You have lost too much," he said, "but you have not lost everything. And you have certainly not lost America, for we will stand with you for as many tomorrows as it takes." Americans seemed to grow closer to the president through the crisis. They'd heard him say he would feel their pain, but suddenly, there he was, doing it, and Americans responded, grateful that he had led us through this national mourning, this terrible time when we realized the terrorist enemy was not always "them" but sometimes ourselves.

Meanwhile, the security gurus cut Pennsylvania Avenue off to traffic, erecting concrete barriers designed to keep anyone from trucking through with enough explosives to blow us into the sky.

I had hated the blockades. They looked ugly to me. Slapdash. Something done fast to ensure security while better, long-term designs were being drawn up. Yet two years later I saw no change. Just concrete reminders that this building was always in someone's bull's-eye.

"Stacy—" I heard a man's voice call my name. I looked up and saw a uniformed guard through the glass, one who knew me well enough to know that I went by my middle name, no matter if it said "Eleanor S. Parker" in the system.

"Hey, there!" I said, entering the air-conditioning and breaking into a huge smile.

"Where have you been?" he asked.

"I was in England, going to school. But now I'm going to be staff!" I said.

"Well, welcome back," he said.

Oh, those uniformed agents with their smiles. They still made me feel lucky and shined on, knowing how quickly their smiles returned to the game faces they showed the general public, given how dangerous their post could be, how this bit of Pennsylvania Avenue—still—was a strange fly strip attracting the world's crazies. There were three instances I remembered from my intern days when someone had charged the grounds: once with an assault rifle, once with a knife, and once with a stolen Cessna 150, crashing into the South Lawn. Each time, these men and women had been the first line of defense, leaping into action.

The second agent checked my bag, then handed it back. "Congrats," he said as the other agent handed me my visitor's badge. I felt the warmth of that first hug back into the fold, forgetting how weary I had been on the crowded orange line, afraid that I had promised Paul Begala that I was something I was not: knowledgeable enough about the nuts and bolts of this place to help him the way an executive assistant must, if I were really going to help him at all.

"I want someone with experience, someone who knows the place," Paul had said. "Someone I feel comfortable around."

We had the comfort part down. When I had worked for George, Paul had been a constant presence, if not in person, then on the phone: bright, happy, always kind, especially that first year. I remembered that no matter how stressful the circumstances, he had *always* kept a good attitude, greeting Heather and me and whoever else was around with friendly smiles.

I respected Paul for his demeanor and his accomplishments. Born in New Jersey but raised in Sugar Land, Texas, the Begala half of Carville and Begala earned acclaim for his work helping to elect Pennsylvania senator Bob Casey in 1986 and 1990, Georgia governor Zell Miller in 1990, and a former adviser to Martin Luther King, Jr., Senator Harris Wofford, in 1991. Like his former campaign strategy partner, James Carville, he did not join the White House staff after the 1992 Clinton win, but he still had his own White House pass, and he advised and wrangled with the crises of the day with the other advisers. When I heard he needed an assistant, I jumped at the chance.

So did my wallet. In August 1997, my professional life stalled, just as my educational future had done a few months before. After the Oxford rejection, I had applied for a paid summer internship at Burson-Marsteller, the international public affairs firm, and was accepted. I moved into the basement apartment of a town house on Hillyer Place, between the Phillips Collection and Dupont Circle, rooming with Laura Capps and others. We lived a sunny, twenty-somethings' Washington dream—except that the rent was $200 more a month than I could afford. Every few weeks I swallowed my pride and called my mother, asking for a few hundred extra, hating myself, feeling that no matter the fancy titles or access I had earned, I would not have arrived in life until I reached a financial independence that, at twenty-three, still eluded me. My internship was approaching completion, and I needed a plan.

Laura Capps told me that Paul was returning to the White House as staff and he needed an assistant. Laura still worked in the White House. I was proud of her; she had gone from being George's intern to being his executive assistant to, now, being a junior speechwriter working in Michael Waldman's shop. I loved Laura. She was as smart as she was gentle, a true beauty inside and out. I listened to her and thought, well, maybe I could follow her path and start out as an assistant—for there need be nothing dead end about that role, look at her! Writing for the president of the United States! And she was only two years older than I.

And I needed the work.

Time-travel back to those "Emerging Leaders" meetings and all the mentor motivational talks, and I don't remember any of our vaunted advisers saying that sometimes we take work because it's the one job available and right in front of us. We do. However, helping Paul wasn't just a job; working for him felt like a natural next step. Unlike at nineteen, when Heather had offered me a job, I now had my degree, and I could spring ahead into other roles when the time was right.

To be staff. The idea resonated deeply inside me still, for didn't every politico intern dream of this, no matter the position? To cross that river between us and them? How many of us watched *The War Room* and cursed fate that we couldn't have dropped everything and moved to Little Rock and been forever part of that sacred circle of the first staff, the most trusted, and often the highest ranking? I couldn't time-travel back to Little Rock, but I could attain a certain completion if I became Paul's assistant.

My decision was made. I called Paul and asked to be considered. Within days, he snapped me up. A few weeks later, in September 1997, I walked up the stone stairs that led me back to the first floor of the Old Executive Office Building.

* * *

"THE ERA OF BIG GOVERNMENT is over," proclaimed President Clinton in his January 1996 State of the Union speech, one year after the seating of the 104th Congress, with the Republicans' net gain of fifty-four seats in the House and eight seats in the Senate. The president's popularity had suffered during his first two years as president, especially after the failure of the administration's health care reform legislation, often characterized as socialized medicine or "Hillarycare." In 1994, the GOP had campaigned on the "Contract with America," a list of promises to rein in what many felt was bloated, runaway government. It had also campaigned hard against President Clinton. After their landslide victories, Republicans claimed they had led a "revolution."

The 1994 election was a massive rebuke to our boss. The voters spoke, and in the West Wing the next morning I felt buried under the rubble caused by an earthquake, as if we did not have the breath yet to move a toe, to see if our spinal cords were still intact. Even as a kid volunteer I felt this. I was in the organism, feeling the same quality of dread I had felt the morning after counsel Vince Foster died.

The day after the 1994 election, I watched George enter and leave the office, a pall over his pale face. Heather kept working as if busy hands could get us through this. This was no time for jokes or talk. Were we even alive? I still did not know. The rubble could be sealing us up, shutting off our oxygen supply. At minimum, the momentum was no longer ours. It was theirs, belonging to the new Republican leadership, the new Republican freshmen, and even to that new Republican governor of Texas, George W. Bush.

Later that night, after my stomach stopped aching and just felt dried and empty, I made a card. Inside, I wrote George a note. I said I was sorry about what happened. Then I included a quote that I attributed to FDR but originating in an ancient King Solomon tale, used to express a timeless truth: "This too shall pass." I slipped the note inside one of my schoolbooks so it wouldn't get bent or smudged.

The next day, I gave George the card. I did it in front of Heather so it wouldn't look weird. He must have opened it promptly, for he came back out to the anteoffice, looked me in the eyes, and said thank you. His eyes said, *Yes, the going's been tough.* Yet something about George's steady ways reassured me that we were going to survive.

And we did. In 1996, President Clinton won reelection with 49 percent of the popular vote, winning 389 electoral votes to Senator Bob Dole's 159. He did so by promising to be the best steward as we crossed "that bridge to the twenty-first century," and by touting his achievements: the Family and Medical Leave Act, which mandated that employees of large firms be allowed to take unpaid sick or pregnancy leave without having to fear termination; the Brady Act, which mandated a five-day waiting period for handgun purchases; the Deficit Reduction Act, which dramatically reduced our national debt; the North American Free Trade Agreement (NAFTA), which opened our borders to free trade; the crime bill, which was supposed to put 100,000 more officers on the street; ending welfare "as we know it" with welfare-to-work initiatives; and the creation of AmeriCorps, which oversaw organized volunteer efforts and rewarded students with stipends and scholarships. None of the victories had been easy—every single one had required the coddling and arm-twisting of congressmen and -women and often protracted, dramatic negotiations that sometimes led to tie votes in the Senate broken by Vice President Al Gore. The Clinton administration may not have provided everyone with health care, and the president's improvements were often called incremental, but these victories all felt hard-won, and administration officials were proud to provide tangible deliverables to real Americans.

A midterm election might reproach the incumbent, but reelection provided vindication, proof that the first time around had been no fluke.

EVEN IN THE DIM HALL I recognized the back of his head, the cut of his suit coat, his walk. Deliberate. The civil servant who had predated us and would surely outlast us, too. I hurried up so that I was just behind him.

"Mr. Good!" I whispered. The man turned around.

"Why, Stacy," Terry Good said. He looked at me for a long moment, his kind eyes smiling. "Welcome back!" I felt the rush of affection inside me like hot blood to my skin. I hugged him. My new life could wait a few minutes while I spoke with Mr. Good, the director of records management.

Ever since I had begun handling George's mail, my work routine had involved delivering documents to Terry Good's office. Any document or correspondence created or received was to be sent to the Office of Records Management for archiving. Nothing was to be destroyed. If an original could not be sent, a copy of the original had to be. For me, this meant taking our bin of responses—copies of such papers clipped to the original letter—to Mr. Good's office in the gray light of the OEOB. These bins of papers were our legacy. As an intern, I was too young and idealistic to know that others saw these texts as something else: a paper trail. We processed clues for historians and investigators alike.

Mr. Good would close the door. "How are you, Stacy? How are your studies?" I would tell him, and he would listen. He made the time. His work life may have been the preservation of texts, but he remained willing to listen, human to human, if only for ten rejuvenating minutes.

Terry Good told me about how Washington used to be. How at cocktail parties, if a stranger had asked you about your occupation and you replied that you worked for the government, decorum dictated that you dug no further: your work was confidential and should be respected as such. A decorum that did not exist like this anymore in this nosy, pushy time, where newspapers splashed people's personal lives across their pages. I mourned the passing of a politer time but also understood that perhaps this was the price of our greater egalitarianism; more of us were less cowed by authority just for authority's sake and we felt safe to speak to power, if only to ask questions. But I never wanted to be the person who disrespected a man as graceful as Terry Good by prodding where I wasn't welcome.

Mr. Good led me into the familiar Records Management suite. Ah, the same steady activity: the men and women working in the carrels. Civil servants, mostly black, reading and sorting documents. I nodded and smiled at those I knew, and they smiled back.

We walked into Mr. Good's office. The room was filled with bright morning light, the kind I associated with early school periods. He closed the door and offered me a seat.

I told him about England. "I kept up," Mr. Good admitted, telling me he had seen the cards I had sent to President Clinton and Rahm Emanuel. I laughed, having forgotten how something as quickly tossed off and seemingly forgettable as holiday cards or "this is how I'm doing" notes became part of the public record— and channeled through Mr. Good's shop—if you mailed them to the White House.

He asked me about my campaign advance work of the summer before, knowing that I had left Washington to go on the road. "They threw me in, like they always do," I said.

"I'm sure you did fine," he answered. I told him how scary my first advance assignment had been, how I had felt like a beginning swimmer dumped into deep water. "Look how you did with that correspondence operation," Mr. Good reassured me, and that made me smile, because he knew from experience how hard I had worked to manage all of George's mail. Thrown in like a swimmer then, too, even though I'd had help and Heather, who, despite her strict standards, had always had my back.

"But this was different," I countered. "They had me doing the RON, on my very first trip, in Little Rock right after the convention. All they do is send you to a two-day 'advance camp' and give you a manual and phone numbers of other RONs and expect you to perform competently. No shadowing another RON first. Just go out there and do it. Even though your Secret Service and White House Communications Agency counterparts are experienced professionals. I felt sorry for them even when they were rolling me, for how unfair was it that they had to deal with my greenness? So disrespectful to throw us out on the road like that, for in the end, we have the authority, speaking for the president and his wants, and while the counterparts have to show us how everything works and advise us as to the best course, we're still supposed to be in charge. We're still supposed to know enough to lead. We're representing the president!"

Mr. Good smiled, no doubt amused by my earnestness. I could always make him smile. I noticed then that his face seemed wan. Tired. With all the miniscandals and document searches imposed on the Clinton White House, I wondered if the stress of it all was starting to wear on him. Terry Good was the custodian of White House documents. Accusers came knocking.

I FOUND PAUL'S NEW OFFICE: 147 OEOB. I stood in the high-ceilinged hallway near the stairwell that led down to the Seventeenth and G Street ground-floor entrance and knocked on our door. I heard Paul say, "Come in."

"Paul!"

Paul rose, his smile followed by a quick grimace saying that this was not the office either of us had expected the new counselor to the president to be given. He welcomed me and told me we had White House Mess reservations and I was thrilled: though I had ordered takeout hundreds of times, no one had ever taken me to the Navy-run dining room for a sit-down meal. This privilege belonged to senior staff, and they often used it to impress outside guests. "Make yourself at home," said Paul as he showed me to my desk, a heavy, solid thing that nearly matched his, a few feet away. Another young woman had been his interim assistant, and together they had started a system of call sheets and appointment keeping. I would take this over now. I slipped my bag beneath my desk and felt my fingertips on the keyboard, hoping I could be as good for Paul as Heather had been for George.

But the room was surprising. I looked at the paint the color of yellowed White House letterhead and the cracks like lightning racing down from the high crown molding and wondered if they last painted this room in the Bush or in the Reagan administration. Paul would never fuss about real estate, but I wondered if the chief of staff or whoever had made the decision had banked on Paul's good nature, because not only was he exiled to the OEOB, he had to share his single room with his assistant. Me.

The two of us faced the door.

Paul explained that this was temporary, that our West Wing accommodation was not ready yet, for there was no space available unless someone else was evicted from his or hers, and he was not going to make an issue of it. I nodded. That was so Paul Begala. I was concerned for what was rightly his, but he refused to sweat the ego issues. After I sat in my new black chair awhile, I realized that I was the unintended beneficiary: working next to him, I'd have a version of what I had always wanted, the trenches feel of the 1992 War Room. No anteroom kept me in a tight bright box, with Paul behind closeable doors. We were together. I could listen and learn and feel like a team. Maybe this would be okay, I thought. Maybe I could do this, and do it well.

* * *

AND SUDDENLY SPINNING, spinning, back in the White House world as if I had never left, answering Paul's phone, organizing Paul's schedule, providing the grounding and backup that a counselor to the president, running a million miles per hour himself, needed to perform well. The moments blur like a sped-up filmstrip of faces, new, old, cold, and warm. I spoke to Diane, Paul's wife, for the first time and found her to be sweet and down-to-earth—a huge relief, given how hard some wives were known to be on assistants, leaning on them as if they worked for their household. I turned in my new forms for my updated security background check. Five years had passed since my first one, back when I was eighteen. Tell the truth, I thought, so I did, proud of myself for being crystal clear about my past drug use, the question that seemed to flummox every aide I knew, for almost every aide I knew had inhaled. I also paid off an outstanding phone bill from the last time I was in D.C. before submitting the financial disclosure form. Security staff called me down and took a new picture for my brand-new badge, and soon it was just like the old days: lift the badge to the sensor, let it fall back and hang from my neck, walk to wherever I wanted to in the complex.

Just like the old days. Including the same low-level dread of starting new jobs, unsure of how to do them. I learned who to ask for help and who would bite my head off whenever I interrupted them: ask simple questions about meeting schedules, and some assistants could act as if hurling me a lifesaver might pull them into the open water. I called it "twentysomething syndrome": thrown into these glamorous, stressful scenes, some young staffers acted pompous and self-important, more like characters they saw on TV than their authentic selves, as they groped for ways to cope with the workload and the pressure. But other staffers refused to adopt this attitude. Ruby Shamir, for instance, assisted communications director Ann Lewis, and she kept her friendliness; she always took a moment to answer questions. Her help, right when I needed it, had felt nothing short of godliness.

Or just look at Paul. Mention his name to any Democratic staffers who knew him and their eyes lit up, for they knew how good-hearted he is. Paul proved by his success that you could be a contender without being cold.

Still, my colleagues could make me stew inside. Why be such an ass? I'd think, hanging up the phone. Then the phone would ring again and I would see the number on the readout and know I had something so many of those other young assistants did not: relationships. A depth of feeling with bona fide power players that they could not have, since they had been here for months, only a year maybe, and I had started in the White House when I was eighteen. When my favorite journalist called our office for the first time, and I answered, "Hello, this is Stacy," and I heard the joy in his voice, I felt okay.

More than okay.

Close your eyes. Sink down in your leatherette office chair. Take your toes and push your shoes off by the heels and pull your legs into yourself, fetal style, so you're as comfortable as can be. Then answer his blinking red line. Listen to the way he says hello once he hears your voice, once he knows it's you.

I doubt the journalist treated me differently than he did the other assistants and staffers, using his charms and wit to worm his way into our affections and to the top of our call sheets. So what, I thought. I loved his attention. I loved his voice. On the phone, he had President Clinton's storied ability to make you think that you were the only person in the room, the only person worth talking to, if only for a few minutes.

Which made him excellent at his work. Journalists could be ham-handed. They asked questions you're not supposed to answer and did so without finesse. Like a kiss that came in too fast, point-blank questions got an instant rejection. I remember one of the

CNN reporters calling in and asking me the subject of the president's radio address, just asking it straight up. He knew I couldn't reveal anything. I sat there thinking, not everyone would be so artless. The masterful knew how to pivot, to massage, to slow it all down. Lesser journalists just pushed. The best asked how you were, then remembered days later what you'd said. How many high-powered insiders fit that bill?

"Hello there, Stacy. How are you today?"

"Better. Now that you've called."

With the phone to my ear, I was back in the circuit. Take that, rude assistants. Maybe you excluded me from your confidence and your happy hours. But others up the Washington chain did not. White House beat reporters called and bantered, sometimes took me to lunch. Some had become real friends. Those thoughts gave me comfort whenever I felt spurned by my peers during that unsure time.

THE YOUNG, FEMALE SPEECHWRITER entered our office, and though I closed my desk drawer too loudly, no one turned around. I was glad yet convinced that the slam's going unnoticed confirmed my status as the invisible girl in the room. Paul was hosting a rare meeting in our OEOB office, a message meeting for the President's Initiative on Race, Clinton's attempt to facilitate a meaningful national dialogue on prejudice, equal opportunity, and race relations. Even though I was in the room, I was not in the room, for Paul did not invite me to participate; I was his assistant, not one of the junior communications staffers who took directions, then drew up drafts of presidential speeches and statements.

We never held meetings in our office. We sat there every day, the two of us, with no meeting table and few chairs, in a room that was fine for working in but not for holding gatherings or receiving guests you wanted to impress. Regardless, Paul called his race meeting for our office, so there they all were, some with chairs they had brought in from next door offices, no one thinking to complain, because they were junior and Paul was their beloved senior.

The speechwriter looked my way, and my smile tightened. She was an attractive, slim woman, and I knew her only from the halls. How much older than me was she? Two years, maybe three? I had just come back from my scholarship to Oxford. But there I was, answering the phone, while she readied herself to participate.

I tried not to let that bother me.

Just as I tried not to listen to their meeting. It was hard not to, though, for the participants packed our office, their circle pushed back against the walls and our desks to accommodate the newest arrival. Paul looked like the den father, at least fifteen years older than the next-eldest participant. I concentrated on finishing my stack of correspondence, on answering the phone on the first ring.

Of course I heard them all. This was the first question they discussed: Do sports teams intentionally pick more whites for the second string and the bench just to have white faces for the majority of ticket buyers to identify with? The underlying assumption being that if they did not, said the staffer, all their players would be black, because now all the best players are black.

The speechwriter piped up, "This is just like at Berkeley. Where if you went straight by the numbers it would be 87 percent Asian!"

I looked at the speechwriter. I wanted to go pat her on the head and ask her why she was in our office making what I thought were obvious points about race—points I could have made myself, given the chance.

I took a deep breath and circled words on my notepad, over and over, until the scratches obliterated the words. Why be mad at the speechwriter? I chastised myself. She was only connecting with the question in an honest way.

But look around, I thought. Who's the only black girl here? Me! Have I been included? No! On the one subject where we junior staffers could make as many valid contributions as everyone else—because no one in that room, besides maybe Paul, could claim any expertise on race, sports, or town halls.

The big joke, of course, was that despite my rising indignation, I would have been equally offended if they had tried to pull me on this—what, because I'm black I automatically know what it's like to be Michael Jordan? The only time you include me is when it's about race? Yet when Paul had announced the people present into the speakerphone, he had excluded me, and it stung.

I clicked to the e-mail, then to the phone log, then back to the letter draft as an energy rose in me, dark and rustling. This was the same feeling I'd had when salaries were published and I discovered another assistant made $2,000 more than me. Or someone my age had a more substantive job. Anger over the kinds of things I used to think I was so above, before. Back when I wasn't staff. Back when I didn't feel stuck.

It was easy to forget that a junior communications job was within reach. Laura Capps had gone from being George's assistant to working in Michael Waldman's speechwriting shop. There was no reason that I couldn't have a future there, presuming Michael didn't hold that forgotten hotel room in Little Rock against me, which I was sure he did not.

It was also easy to forget that the talented speechwriter in front of me had reached for what she wanted, and now it was hers. Back then, we didn't use the term "hater," short for "player hater." But in those jealous moments, that's what I was, for sure, and I hated myself for it.

The problem was, I hadn't had my job three months and I felt as though I were spinning my wheels. We worked in a hierarchy, and I was support staff. Every morning I sat in my chair and worked on Paul's schedule, opened mail from constituents, wrote his responses, answered his calls and put them through. That's what I did. I worked within set bounds. I could not be handed bigger and better projects because Paul *was* the project. Expansion could come only through promotion or resignation.

I'd always known I didn't want to do support staff work. I just hadn't known that I would chafe so quickly.

11

Family Reunions

I DID NOT MEET MY MOTHER'S FAMILY UNTIL I WAS TWENTY-THREE years old. A quarter century after my mother left Kansas to begin a new life and never returned home, her cousin Donna had tracked her down using a 1-800 search service. My mother returned from work one day to hear Donna's voice on the answering machine.

My mother called me at work in Washington. "Oh, my God," I responded. "Are you serious? How are they doing?" My mother said she didn't know; she hadn't called Donna back. In fact, she hadn't called any of her family in all those years. We're talking radio silence. She had always liked this cousin, but was it a good idea to rip open what had scarred over so long ago? Neither of us knew.

THERE IS NO ME without my German ancestors from Victoria, Kansas, but for my whole life, they were simply names and roles, characters from a book on a shelf that my mother kept closed. I learned the word "estranged" early and kept with my mother's unspoken wish: leave it alone. My mother looked ahead, not behind.

My mother, Carol, was born in 1950, the first child of Elmer and Irene, and would be the oldest of five siblings. The nearest big town was Hays, as in Fort Hays, the once violent frontier outpost where "Wild Bill" Hickok had served a spell as mayor. According to the town's Web site, Victoria was founded in 1873 by thirty-eight Scottish immigrants. They were soon followed by Catholic "Volga-Germans" fleeing conscription into the czar's army.

The Kansas around Kansas City is hilly, green, and welcoming. But Victoria's plains stretch for mile after treeless mile, windblown and flat as a frozen pond. The Scots mostly gave up on the unforgiving land, but the Germans stayed on, using ingenuity and toughness to reap what they could, quarrying stone to build St. Fidelis Catholic Church, the Cathedral of the Plains. When completed in 1911, St. Fidelis was the largest church west of the Mississippi. My mother attended this church as a girl, sitting with her family in a back pew. She recounts being hungry as they waited for communion and unable to reconcile the piety at church with the meanness around her.

Her father, Elmer, served in the Pacific theater of World War II and, when he returned, married Irene. He had fought in brutal island fights and spoke little of the war. Elmer and Irene grew alfalfa and wheat and kept cows. They raised their family in a compact frame house that could be so cold on winter nights my mother's perfume bottle would freeze over.

Neighbors found oil on their land; Elmer did not. Neighbor girls wore crinolines under their skirts; my mother did not. Instead, my grandfather drank. He started fights. He eventually stopped drinking, and this meant that her younger sisters Wanda and Mary grew up in a saner, safer household. But Carol and Ron, the eldest sister and brother, grew up learning to cope and storing up anger against a man who could explode at any moment.

As I grew up, my mother rarely spoke of her father, and today I marvel at how easily absence can go unquestioned. I knew my mother had a family in Kansas. I knew she would not see them; therefore neither would I. This felt like a fact of life. And since no one gave me direct explanations, I made up my own. I assumed Elmer had disowned her for marrying a black man. It was not until Donna's call that my mother explained that she had stayed away of her own accord, that her anger at her father for the way he had acted during her childhood had fueled her desire to keep her life separate from theirs, even if it meant cutting off contact with the rest of her family, including her mother, whom she loved deeply.

Growing up, I had never questioned what felt unchangeable: we lived in Michigan and they lived in Kansas and a gulf lay between us, too deep to cross.

MY MOTHER PICKED UP the phone and called her cousin. Delight and happiness, for her, for Donna. Then my mother called her parents. I could only imagine how heart-stopping those first moments must have been. But they spoke warmly to each other. Welcoming. They to her, she to them. Carol told them of her daughters' accomplishments, of her comfortable, happy life in Troy. Soon came photos, letters that she shared with me—a whirlwind of history in our hands. I marveled at my cousins, girls with big brown eyes just like mine. A blue-eyed boy cousin who looked like Ramsey, my former roommate and crush. Uncanny. How magic and strange these reunions felt, if only in pictures.

Within weeks, my family made our first trip to Victoria, Kansas. I flew from Washington, D.C., and met them in the Hays airport. In the waiting room they stood: my aunts and uncles and my grandparents of German descent, Elmer and Irene.

Irene hugged me sweetly. She had short white hair and a kind face, just like my mother's. Then Elmer hugged me. He was now old, with replaced hips and clipped white hair. He was warm. Friendly. Not the crew-cutted brute in a T-shirt I imagined him to be. He bonded with my stepfather, talking World War II with the émigré Filipino doctor, apparently telling stories that no one in the family had heard before.

My mother remained her even-keeled self. She found herself confronted with aged parents, and recriminations did not seem in order. They were the same people but not the same people. She focused on the positive, on reconnecting with her mother, her siblings, her extended family. Time changed so much.

But not everything. I have now been to my mother's birthplace twice, and each time I was overwhelmed with feelings of loneliness, of being trapped in a house you couldn't escape. Go outside for air, and you're surrounded by country so flat that the wind races along and strips off all the humming, hovering energy necessary to collide and combust—the pixie dust that breaks ideas through the psychic barrier and into reality, that makes city centers rich in possibility. The land felt as dried up as the moon. Even when wheat and sunflowers showed proof that this treeless land could produce, I still did not believe it. My instinct told me that the only good idea born here was to run.

Just as my mother did in 1971, when, at the age of twenty-one, she enlisted in the U.S. Army. Carol didn't love or even like the military life, but as long as the army didn't send her back to Kansas, she was okay. After Alabama and Texas, they sent my mother to Fort Sheridan, Illinois, outside Chicago. It was there that she met Edward Stanley Parker, one of the fresh vets waiting to be discharged. Stan, aka Butch, was a talker. Charismatic. Brooding. They were two very different people, and a connection grew.

My mother discovered she was pregnant with me, a little girl whose father was tall, big-shouldered: as American as you can get with brown cream-of-wheat skin, the mixing already done. Now, as an adult, I understand that these situations don't always end in birth. My white mother must have wanted me, and for this I am utterly grateful. My black father must have wanted me, too, for he did not promise my mother love if she made me disappear. My father charmed my mother in a different way. "Come to Detroit," he said. "People will help us out. Let's go."

Funny, I had always felt haunted by my family, by the absences on both sides. But once presented with the chance to make deeper connections with my mother's family, I faltered, too swamped by my White House work and the nagging feeling that I had extended family in Detroit who deserved my loving attention, that they should not be neglected because the other side of the family tree was suddenly visible. I kept up with my Kansas family and would later spend time with my aunts and nieces, so full of life and verve that they felt "mine" from the very first hugs. But it was my Detroit family, especially my father, who remained on my mind, remained the mystery to me that I struggled to solve.

When Butch died, my mother remarried and we left the city for the suburbs. I stayed close to my grandmother and saw her often. But I saw my extended relatives only at funerals and occasional holidays. At the 2006 funeral of the matriarch of my father's family, Aunt Thelma, I entered an unfamiliar West Side Detroit church and encountered relatives I hadn't seen in years. I approached the rising matriarch, Thelma's daughter Brenda. She had ten children of her own. I thought of her as mythic, as if she led her own tribe.

Brenda placed her soft hand to my face. "You look just like Butch," she said. She looked unsettled yet dreamy, as if I'd taken her back thirty years. As if I'd brought the dead back to life.

I smiled, demurely. Matriarchs can have such power that they inspire obedience even in moments like these. She placed her printed scarf around my neck, saying that she had never been able to give me anything. I thanked her. I hugged her. I kissed her cheek.

But I hurt inside. Brenda had a lifetime of memories of her Butch, picture after picture in her head of his face, his smile. Imprints of a man who will always be unknown to me. He didn't live long enough for me to get to know him. And it didn't matter how loving I was or how cute I was or how much I wanted to be with him; there was nothing his girl child could do to convince him to stay.

So I KEPT WORKING. The funny thing about working in the highest echelons of government is that you can dive so deep into the problems of the world that you can escape your own problems—or at least try to. You can go fully underwater and stay there until you are so waterlogged that you feel no pain. Family trauma feels far away, like something you left on the shore. You can work sixty- to eighty-hour weeks, and since that's normal, your family may excuse you from day-to-day duties, may give you a pass, because you're doing Important Work. You may be on TV, but you're checking out right before their eyes.

When it's Important Work, no one dares call it avoidance.

I wonder how many choose the way of power, the lifestyles all around me, so that the wounds inside them can stay untouched. Or how many of my bosses, even the Clintons themselves, found their way to this life so that they would never have to air those wounds. Wounds I think of as unlined tanks of benzene that leak into the soil and then into the water table. How if we ignore these toxic places, we lose the chance to cleanse and heal them before the hurt turns malignant.

Soon, a president's actions that were no doubt rooted in pain would be pushed to the forefront of the collective consciousness. Not just of his White House staff but of the nation and the world.

12

Cliffside at the Abyss

I RUSHED INTO THE OFFICE, AND I KNEW TO GO TO DRUDGE immediately. This was Sunday morning, January 18, 1998, and the phone rang and rang as I sat in my coat and answered a line and booted up my computer. I didn't usually work on Sundays. Paul held his receiver to his face, staring at Matt Drudge's telltale black type across his computer screen. He looked at me and tried to smile, but it was a foxhole smile.

"This is all vicious rumor," he said. The phone kept ringing and it wasn't even 8:00 a.m. ABC's Mark Halperin. NBC's Tim Russert. CNN's John King. Journalist after journalist, and I took their calls. Paul took their calls, but the phone log filled right back up.

"The independent counsel . . . no, he won't resign . . ."

I entered "www.drudgereport.com" into the browser and clicked where millions of people had to be clicking at once, onto this story:

BLOCKBUSTER REPORT: 23-YEAR OLD, FORMER WHITE
HOUSE INTERN, SEX RELATIONSHIP WITH PRESIDENT
** World Exclusive**
Must Credit the DRUDGE REPORT

At the last minute, at 6 p.m. on Saturday evening, NEWSWEEK magazine killed a story that was destined to shake official Washington to its foundation: A White House intern carried on a sexual affair with the President of the United States!

Drudge's account provided official Washington—and everyone else with a computer—with the first details of the Lewinsky affair, including the allegation that the president and the intern had sexual encounters in the Oval Office study. In addition, Drudge spotlighted journalist Michael Isikoff's behind-the-scenes struggle to get the story published, and the "blind chaos" at other news organizations forced to play catch-up with the most nuclear scoop of the Clinton presidency. He ended the piece with: "The White House was busy checking the DRUDGE REPORT for details."

We were. From the minute I read the headline, I believed everything. No moment of doubt. No moment of questioning the motives of the accusers or reporters. The president had clearly interacted with this young woman; to what degree I did not know. When we learned of Vernon Jordan's supposed involvement, the allegation that Mister-Fix-It Extraordinaire had picked up the phone and helped his friend get the girl out of town with a job in New York, I never doubted the charge. That seemed pure Jordan. Nothing necessarily sinister about it either: help for a friend who needed an obsessive girl happy and out of the way. Something he could do in two seconds. Something he'd do for so many. POTUS wouldn't have to ask; some things were understood.

I had no proof. Just a gut feeling, based on how I'd observed these powerful men over the last five years. And the fact that the president's priors ran a mile long.

Since the Gennifer Flowers scandal, I had always given the president the benefit of the doubt, siding with him over his accusers, tacitly agreeing that if the women's motives could be questioned, so could the charges. Not anymore. I stared at Drudge's words and knew that this was the tipping point.

I took the calls, typed them in, took the calls. Paul returned them one after the other. The pit in my stomach felt like a cannonball. We may have been capsized. We may have already drowned.

Senior staff called meetings. The Roosevelt Room. The Oval Office. In nooks unseen. The phone kept ringing, ringing. Soon the news anchors called. They never called. They wanted Paul. They wanted answers. They asked me, "Monica Lewinsky, do you know her?" "Well, yes, I knew her. Not well. But I knew her." I answered once before I knew I needed to shut up. Before I went on the record, too.

The calls continued. The meetings. The scrambling. Defense, when the advisers should be on offense. The State of the Union just days away, and everything should be pushing toward this speech. Instead . . .

"He looked me in the eye," Paul repeated, "and said he didn't do it. When the president of the United States looks you straight in the eye and tells you something, you believe it."

I didn't. I sat there working, but seething inside. I was so angry at the president. How could he allow his sex drive or, really, the gaping maw inside him, to derail his presidency? Staffers galvanized their talents, their energies, their everything in service of the president's dreams. Of the shared dream . . .

Later, I would feel sympathy for the president. The older I got, the more I understood how our desires make us do stupid, reckless things. But back then I kept my thoughts to myself and kept answering the phones. Paul and I didn't talk about it. Behind that pained smile, I felt an apology. *I'm sorry for dragging you into this.* For none of us could see the future. We had no idea how this would end.

* * *

WHAT HAPPENED?/ABUSE OF POWER!/Oral sex!/Obstruction of justice./ Unfit to lead./She was a stalker./Are there others?/How could he!/Moral degeneration./Unfit to lead./What kind of man is this?/Is he going to resign?/ If these allegations are true, he should resign!/Gifts, there are gifts./How do we explain this to our kids?

Which scale can measure this force? Richter? Saffir-Simpson? We're talking the power of every single American trained on 1600 Pennsylvania Avenue in anger, in disbelief, in ridicule. The fast, loud insistence that the president had been *found out*, that he had proved his detractors right by revealing his *moral corruption* for the world to see and judge, that he should be *ashamed*, that we're in *crisis*. If there was a weak bone in the president's body, if he was prone to panic or shutting down completely—if he could not hold it together somehow—I could only imagine that he would have been obliterated by this force.

Instead, on January 26, President Clinton stood in the Roosevelt Room, wagged his finger at the camera, and declared, "I did not have sexual relations with that woman, Miss Lewinsky. I never told anyone to lie, not a single time. These allegations are false, and I need to go back to work for the American people."

I watched on our office TV. Every channel had to be covering this. I said nothing, just watched. Then the phones went nuts.

"No, he won't resign!" Paul scoffed when journalists asked him, more than once, if that was the plan. "Didn't you hear what he said? These allegations are untrue." Paul would never crack, but I could tell he was exasperated. Later I caught him joking with Mark Halperin about working for "President Gore."

LEGALLY, THE ORIGINS of these bombshell revelations lay in the Paula Jones case. In 1997's *Clinton v. Jones*, the Supreme Court decided that Jones could pursue her sexual harassment civil case against the sitting president. Pretrial discovery began, and Jones's attorneys subpoenaed those they felt could corroborate the fact that Clinton had propositioned women who were not his wife. One of those women was Monica Lewinsky.

In June 1995, at the age of twenty-one, Monica Lewinsky joined then Chief of Staff Leon Panetta's office as an unpaid intern. Monica was raised in Beverly Hills; her internship was arranged by a connected family friend, Walter Kaye. She worked primarily in the OEOB but sometimes in Panetta's West Wing office, covering phones when staff assistants needed relief. Monica may have come from affluence, but she displayed signs of deep emotional neediness: her mother later spoke of Monica's extreme distress and suicidal tendencies after she and her father had divorced. Despite the early pain, Monica cut a happy, eager figure as she walked the West Wing halls, the same halls we would later learn had been the site of her first dramatic flirtations with the president: during the November 1995 government shutdown, when only essential staff and unpaid interns could report to work, she had flashed her thong underwear at him. She later saw the president alone, in George's office, when none of us were there, and he beckoned her to come in. They walked through George's back door to the windowless hallway adjacent to the study and shared their first kiss. Later that night, he invited her back to George's office, where they slipped into the private study. It was there they had their first sexual encounter.

Soon they had more rendezvous, but always of the quick, hidden kind. He called her at home, in the middle of the night, or right as the sun rose—any stolen moment he could seize—and they talked dirty to each other. Reading the play-by-play of their relationship, as anyone may do in the publicly released Starr Report, he apparently was very hot/cold with her and often said he needed to be "good" and would cut off their relationship only to give in and resume the phone calls. And the gifts. And the moments in the study, with the door open and the heart-stopping knowledge that at any moment they could be discovered *in flagrante*. According to Lewinsky's testimony, she and the president had a total of ten sexual encounters, all of them in or near the study just off the Oval Office and, on occasion, in the Oval Office itself.

Monica fell in love with this married man. Or worse, she fell into obsession. I once read of an experiment with guinea pigs that showed that if food is permanently withheld, a guinea pig adjusts—she finds food elsewhere or dies. If food is given at regular intervals, the guinea pig adjusts to this as well, staying regular and content. But if food is given haphazardly, the guinea pig goes crazy, obsessing over the feeding tube, going back repeatedly because at any moment the lifesaving nourishment could return.

The more Monica cruised the Oval Office and presidential events looking for the president's attention, the more staffers noticed, including Deputy Chief of Staff Evelyn Lieberman. By this point, Monica had secured a low-level staff position in the Office of Legislative Affairs. Lieberman, no doubt wary of Monica's judgment and of the president's inability to withstand temptation, had Monica reassigned to the Pentagon.

The move pushed her straight into the bosomy intimacy of a new confidante, longtime government secretary Linda Tripp. Tripp was a Bush White House holdover who herself resented having been transferred to the Pentagon and was more than willing to listen to Monica complain about her unfair treatment. Their talks turned into the most incredible dish sessions when Monica revealed the reason for her banishment: her ongoing affair with the president of the United States. Dish that Linda Tripp, with a book deal on her mind, soon began to audiotape. Dish that Tripp eventually fed to Whitewater Independent Counsel Kenneth Starr in time for FBI agents and U.S. attorneys to detain Monica Lewinsky on January 16 in a Ritz-Carlton hotel room long enough to keep her from informing the president that Starr suspected them—allowing a perjury trap to be set for the president's January 17 deposition in the Paula Jones suit. If the president had known they had Monica in that hotel room, he might have answered the questions differently.

Then Drudge broke the story and the world did a double take. *Did he really do this? If so, what moral degeneracy!* Or, *how dare Starr stalk the president? What does his sex life have to do with Whitewater, or with any of us, for that matter?* Democrats like James Carville declared that this was war, that the Democrats would fight back against the Investigation Machine, which had become an obscenity unto itself. Around the world, people talked about the American president and his intern. The French shook their heads. *Look at François Mitterrand's funeral. Both the wife and the mistress were there.* The Russians shook their heads. *How lucky we would be if Boris Yeltsin had the life force to chase women this way.* Day by day, salacious details dripped and flowed to an international audience that drank it up, while others covered their eyes, screaming, "We don't want to hear it anymore!," then turned up the sound on the TV anytime there was an update. Would we watch an American presidency disintegrate before our eyes? Stay tuned.

I MIGHT NOT HAVE LOVED my job, but I liked my boss, Paul, very much, and we had established our routines. For ten hours a day I had answered his mail and his phone and managed his schedule. I listened to him work his phone. Paul had returned to the White House so he could advise on policy issues as well as political ones. Now, days before the State of the Union, when all energies should have been focused on the policy initiatives to be featured in the speech, Paul had to field question after question about the president's sex life and what he possibly did to keep others from finding out the facts.

My heart broke for Paul. I sensed his anger wasn't just "How could he do this to us?" but "How could he do this to her?" Paul was a family man. Paul was a spiritual man. Paul had a moral code that did not simply write this off as male license or male frailty. If the allegations were true, these violations of trust—these violations, period—bothered Paul at a deep level. And I was sorry that he was stuck defending what must be, in his eyes, indefensible.

ON TV, I watched the stakeout at Monica's Watergate apartment. I watched her get into the sedan. Her hair shone like sable in the camera lights. She looked pretty. At least, at the very least, she looked good for her gauntlet walk.

The nation laughed and condemned her from afar. But Monica was no pop singer, no performer caught with her skirt up. She was a civilian. Yes, she had acted indiscreetly, sharing information about her affair with friends, but she had also been betrayed by one of those wire-wearing friends, Linda Tripp, in spectacular fashion, and was now being coerced to share the most private of sexual details with the independent counsel and, therefore, the world.

Refuse to cooperate? Face conviction. Not just you, but your mother, too. According to Lewinksy's later account, Starr's agents and attorneys had threatened her with twenty-seven years in prison when they'd had her in that Ritz-Carlton hotel room; this after they had surrounded her at the Pentagon City mall food court, to which Tripp had lured her for that purpose, and they'd told her to follow them upstairs. They'd denied Monica the chance to call her lawyer—finally acceding after 5:00 p.m., once the lawyer's switchboard calls would no longer be answered. Monica had refused to answer their questions that night. Instead, Monica had asked for her mother to join her and she did. She retained counsel William Ginsberg. Soon, Monica sat secluded in her Watergate apartment while Ginsberg worked the TV news shows, telling Starr's team that Lewinsky would "tell all" in exchange for immunity. Discussions continued, halted, continued, then stalled.

In the meantime, Starr called other witnesses to his grand jury. For the next seven months, Starr forced White House staff and others to testify, including Monica's mother, Marcia Lewis; the president's friend Vernon Jordan; the president's adviser Sidney Blumenthal; an ex–White House volunteer, Kathleen Willey; White House aides Marsha Scott and Nancy Hernreich; the president's diarist, Janis Kearney; the president's personal secretary, Betty Currie; the informant, Linda Tripp; Special Agent in Charge Larry Cockell; the president's friend and TV producer Harry Thomason; and the president's adviser Dick Morris.

It was not until July 28 that Lewinsky's new set of high-powered D.C. attorneys, Plato Cacheris and Jacob Stein, announced that they had brokered full immunity for her and her parents, leading to her first grand jury appearance on August 6. As part of the agreement, Lewinsky surrendered what became known simply as "the blue dress"—the dress that contained a stain that was proven to be President Clinton's semen, the dress that Linda Tripp had encouraged her to keep safe and never clean, the dress that Monica was roundly criticized for keeping, even though her action made me think that consciously or subconsciously, she knew that someday she might be called a liar and would need irrefutable proof.

So Monica cooperated. It took months, but she did it. But even if she had cooperated on day one, could she have prevented the subsequent character assassination?

I doubt it. The jokes were too easy. Both the professionals and the amateurs e-mailed and called one another and broadcast their latest. *I didn't tell her to lie in the deposition, I was telling her to "lie in that position."* . . . *What's the difference between President Clinton and the* Titanic? *Only 1,500 people went down on the* Titanic. . . . *What is Bill Clinton's favorite federal program? Head Start.* . . . *Why doesn't Clinton play his saxophone? He plays with his whore Monica.*

Then there were fat jokes. She inspired a special kind of contempt for not being a skinny model or an actress with a look we associated with American beauty queens. Her weight fluctuated, doing so before our eyes. You can look online in 2009 and still find "candid videos of Monica Lewinsky's fat ass."

If Supreme Court Justice Clarence Thomas suffered a "high-tech lynching," this was high-tech pillorying, with Monica's face and hands locked in the town square post. By day, everything foul and rotten was thrown at her. By night, the scene became one big circle jerk. A nation picked one woman to humiliate above all others, and that was she.

I felt as if I were cliffside at the abyss. I too was there during the government shutdown, back when Monica and the president first met. I was an intern in the West Wing. A girl playing adult, just like her.

IN THE FALL OF 1995, the president and Congress played their budget brinkmanship games, forcing a shutdown of the federal government. Twice. Later, these battles would be seen to mark President Clinton's comeback after the harsh reprimand of the 1994 elections, as Newt Gingrich and the Republicans looked too obstructionist and too willing to deplete popular and relied-upon entitlement programs. In the meantime, only essential staff could report to work. Senior staff could show up and do their jobs, but no junior staff could help them. That meant there was no one to do their schedules or answer their phones.

The solution? Let the interns play staff. In several offices, interns subbed for their junior staff bosses. I was one of them, subbing for Laura in George's office.

The West Wing changed with fewer people around. More silence. More empty hallways. When people passed, they said hello to me. They smiled. Maybe they came in and talked a bit. I was a young woman, alone, and there was an intimacy. An easiness. What before had seemed like public display in front of the others now felt like private conversation.

This was heady stuff. I did all the work, yes, but I was also receiving all the attention.

By late afternoon, I felt good. No one had asked me anything I couldn't answer. I'd finished George's schedule for the next day—a task that used to seem Herculean when Heather did it. The phone kept ringing, but not crazily. I even had nice chats with my favorite journalists. I loved being able to talk to them without Laura or another intern nearby.

I overheard someone say that pizza had been ordered. I didn't think much of it.

Usually I sat in the specially created intern nook, which faced the computer and the wall. I never got to see who walked past in the hallway. It could have been Bruce Springsteen or Bono or the president himself, but I'd never know because I couldn't see! I'd miss the chance for a smile, a hello, a chance conversation. Instead I would see only the looks on Laura's face, and I'd have to whisper, "Who was it?" and be satisfied with her reflected glow.

Now I was in her chair, facing the open door. Anyone who came into the office came through that way. Anyone except for one man.

"Do you want some pizza?" the president asked.

I swiveled. Oh my God: the president stood right above me. He was always this strange vision, tall and substantial, his face full of color. I was so used to seeing him on TV that when he showed up like this, I was stunned. Alert. Happy. He had come in from the pantry. George's office was the only one with a direct connection from the Oval.

"I'm sorry?" I said, looking up at him, with an earnest smile.

"Would you like some pizza? I don't know what kind they ordered. I think there's pepperoni, meat lover's, a veggie . . ."

The president was taking the time to explain this to me?

"Sure," I said. I mean, what else do you say? Even if I'd just eaten steak and potatoes, the *president* had just offered me food.

"What kind?" he asked.

I don't even know what I said.

"Okay." He was off.

Was the president of the United States getting me pizza? Maybe I was supposed to follow him. The phone rang. It was Halperin, looking for George. I asked him to hold. I thought I was supposed to be following POTUS. There was no way the president was going to go get food for me—

The president returned. He placed the paper plate on my desk.

"Thank you," I said, smiling.

"You're welcome," he said. Then he left.

I was thrilled that he had been thoughtful enough to ask if I wanted some food. And then, on top of that, to actually go get me the food. Precious few staffers displayed such a caregiving streak. I knew this firsthand; I was an intern. You could tell a lot about a person by the way he treated interns, how he treated those of us low in the hierarchy.

I went back to work. Then I went home. I told my roommates the pizza story. These guys were my old high school friends from Troy, Michigan, Ramsey and Jeffrey, and they kidded me like crazy. "You want some pizza?" Ramsey drawled. Over and over. Both of them laughing. They thought he was coming on to me. I insisted he wasn't. I insisted that he was just being kind and human. They still thought it was hilarious, but I didn't care. It was just a story that made them laugh and made me feel good, because in the end, I had been, for a few fleeting moments, the object of the president's attention.

I didn't think much more about that story after that. "Pizza night" was no turning point in my life. No major epiphany. Only later, in January 1998, when everything hit the fan did the story take on new significance.

For on the other side of my wall sat Monica. She sat outside the chief of staff's office, doing the same kind of junior staff duty. I remember seeing glimpses of her when she walked past our open door. I thought she was cute, with a beautiful smile and a big, effervescent personality. I said hello to her, as many of us young women did when we saw each other. It was not until 1998, when the story broke, that I knew they had screwed around that night. By the time I was home glowing about the pizza slice, they had slipped away into the study, via George's office.

Did you know her? Friends and strangers asked me this constantly. They still ask me this. Yes, I knew her. But not well. She was nice. There's no denying that. But we lived so anxiously back then. I knew Monica as a girl who sometimes laughed too loudly. As a girl whose skirts were sometimes a pinch too tight. I knew she had a certain generosity and carefreeness that could come off as overwhelming. Or as threatening, given the tight, self-regulating confines of the West Wing.

I thought she meant well. But . . .

I knew something she had done, something that in my mind— and in the minds of George's other assistants, staff or intern— marked her as a problem. She had done something that was so "crazy" that all of us knew to keep our distance.

She brought a Starbucks latte to our office for George, unbidden. Just showed up and told Laura she thought George might appreciate the refreshment. Laura accepted the latte in the anteoffice, but that's as far as the coffee delivery went. I presume the drink met its end in the sink.

Laugh at us if you want. But to this girl troll in Gap slacks, Monica's action was a major breach of protocol. Here was an intern, who "didn't even know him," being as bold as I'd ever seen anyone be who had submitted to a background check and wore a blue badge. To us, this was a pass. The kind of thing no one we knew would do. We were mortified.

Monica developed a reputation beyond our office. After the story broke, I heard the *c* word used—the kiss of death for an aspiring young staffer. She was a "clutch," they said. To be a clutch meant that you were so blinded by desire to be near the principal that you didn't care who saw and you didn't let gossip slow you down. In my experience, clutches tended to be interns, junior advance people, and staffers who did not get regular access to the West Wing. Also POTUS friends and associates, usually from the old days, who didn't understand that his time was limited, who thought that their friendship was so old and dear that they need not respect his schedule. The president needed to move, and the clutch was still talking. The president had another appointment, and the clutch wanted to go along. The space around the president was as monitored as it got. You noticed. Everyone noticed. In the zero-sum game of POTUS attention, you looked for the person who didn't fit, who grasped for too much.

Monica managed to be wherever the president was, whenever she could. She always vied for eye contact, for attention, for time.

Clutch. The word hurt my heart. It said, "You are them and not us."

LATE JANUARY, and the days bled into each other. Paul left for more meetings, took call after call. The TV was never shut off. We watched updates. Pundits. Stand-ups live on the scene. So many snapshots of Monica. Early ones. Late ones. Her life on parade.

I thought of high school, of my life in Troy, Michigan: suburban middle class, prosperous. My classmates were kids of Catholic whites who had fled the city or the motherland. This was a school where no one showed up pregnant. This was a school where girls never fought.

One day, I was in a far-off part of the school, in the Performing Arts Center hallway. Lunchtime garbage littered the floor. I was fourteen. I was trying to get to class.

I could hear someone behind me.

"Slut," he said.

He said this to the back of my head; then he passed me. He

didn't turn around. But I knew who he was. A senior. John.* A Polish immigrant with a mane of dry blond hair and a deep voice. An accent. He was seventeen but seemed so adult to me, so utterly male, the kind who pushed you around. I didn't flinch. But . . .

He could have thrown me into the wall if he wanted. Or hit me. In that moment he was young, but he wasn't young. Not to a girl just months out of middle school.

I made it to class. My brain kept reeling, thinking about how I hadn't had sex with anyone, how I may have come close on that New Year's Eve date, but he didn't go to Troy High, that I'd basically only screwed around, and with no one in my school, yet . . . did he know? How did he know? Can everyone tell?

Stop and look at me then in my dark eyeliner and teased hair that fell over one eye. Was that the day I wore the black skirt with the silver buckles down the back? I drew attention to myself in sexualized ways. As all the girls did in their makeup and miniskirts. I had only the slightest idea of what I was doing, as I tried to assert an identity and ask for love at the same time, with little understanding of these motivations. I was fourteen. I knew. But I didn't know.

I tried to listen to the teacher's lecture, but I couldn't. One word from John, and two thousand watts of fear soaked through my fascia on its way to the bone.

Monica Lewinsky went to high school in Beverly Hills. It was a different culture with different problems. But some problems affect us all. Girl to girl, I know that I cannot be fundamentally different from Monica. On the night of the West Wing "pizza party," she ended up being intimate with the president. I did no such thing. Still, we both hungered for male affection and attention. We both knew what it was like to get the wrong kind back.

Just like any girl on Earth. Any girl at all.

* This name has been changed.

CLIFFSIDE AT THE ABYSS   191

DURING THOSE DAYS leading up to the State of the Union speech, journalists kept asking Paul whether or not the president would resign. Paul rebuffed the questions. He steeled his voice. He was indignant. Angry, even. How dare they even ask?

Yet when we sat in Chief of Staff Erskine Bowles's outer office that Thursday night, with so many senior staff there, our eyes trained on the TV above our heads, we knew anything was possible.

The president entered the chamber. He walked slowly down the aisle, shaking hands with all the congresspeople and senators, hugging, smiling, and nodding solemnly. We watched. I felt the catch in my throat, my uneasy stomach. We knew there was a SOTU speech in the teleprompter. A real State of the Union with all its proposals and applause lines. The president and his speechwriters and advisers had worked arduously on the speech. It existed. There was no official plan for resignation.

But there was no guarantee. Maybe he heard the village drumbeat louder than he heard his own heart. Maybe his knees would give out. Maybe his conscience would push him to confess, to try to make things right.

I've heard that people commit suicide not because they want death but because they want relief.

That night, the president would do no such thing. He delivered his speech, his State of the Union. And with each word he spoke, he continued to fight the way he knew how to: he kept moving forward, using every maneuver he knew, while the rest of us watched.

Or found ourselves dragged into drama, too.

13

Truth, Lies, and
Background Checks

Spring 1998, and our lives remained all Monica all the time. My boss, Paul, worked long hours with other presidential advisers, shaping press strategy. He also negotiated and tussled with the White House lawyers for the information he and other advisers needed to mount the president's public defense. When the phone rang that spring afternoon and the caller ID said "Counsel's Office," I took no special notice.

Then the voice asked for me. "Please come down to meet with one of the counsel." Was this for Paul? "No, this is for you. Come by this afternoon."

Directions followed. Not to the West Wing office, nor to either of the Lanny offices—the suites of Lanny Breuer or Lanny Davis, the friendly men whom I felt comfortable around, given how often they visited Paul. No. Go to a different suite, one that was practically anonymous on the West Executive Drive side of the

Old Executive Office Building. I'd never even heard of this female counsel before. I must have walked down that corridor at least a thousand times and never known who sat behind those unmarked doors.

When I told Paul about the call, he took the news in stride, without betraying prior knowledge or alarm. Our focus returned to the phones and journalists and the meetings on Paul's schedule. In fact, I did not think too much about my appointment until I was led into the counsel's office that afternoon.

An older man I did not recognize stood next to the female counsel. I don't remember shaking hands as he introduced himself as FBI and asked me to sit on the couch.

The counsel's window looked into a barren interior courtyard, and the sunlight felt weak. We sat surrounded by the same off-white walls, hairline cracking in places like those in my own office, like any other OEOB office with its state-issued wooden desks and low cloth-covered couches. I sat on her couch pressing my knees together, my back straight and my hands in my lap, waiting for them to speak.

"There's a problem with your background check."

Who said this, I don't know. The minute I heard "problem" coupled with "background check," fear flooded my mind like gas through an engine. Across from me was the female counsel in her late thirties or early forties, with no kindness in her voice. What could it be? What could someone have said?

"We have questions about your answers regarding your past drug use. There are discrepancies."

My eyes locked into hers, then his; then I know they darted upward, the way my eyes always do when I'm thrown off by a question and I don't have an instant answer. I told them I had answered my background check questions honestly. My mouth was moving. I was saying more but I don't know what, telling them that I didn't know what they were talking about.

"You need to be honest with us," they said.

I had been! I had written down *everything* about my high school drug use. All of it. A point of personal pride that I did. Down to the half hits, for yes, I had used LSD. I had smoked marijuana, too, but only rarely and in such small amounts that I never got high. And yes, I was proud of my bravery, of my willingness to put it all out there on these forms and not be ruled by my fear of the consequences in a town where, in 1998, prior drug use still scandalized.

"You do realize that giving false statements to investigators is a felony?" the counsel asked. "This could mean jail," added the FBI agent. "At the very least, you could be fired."

But what were they talking about? *I didn't lie!* My nervous system pulsed a hot sensation to my brain that this was fight-or-flight time, but I sat frozen as my mind madly scrolled itself, trying to figure out what they could be talking about. I had no idea.

"We need people of good character. Not those who lie in their background checks," said the agent. "The administration needs staff they can trust."

The words burned me badly. I wasn't used to being looked on suspiciously—in fact, I tried to live my life so no one would ever have cause.

"Do you want to be an embarrassment to the administration?" asked the counsel.

Time stopped. How was it possible that my teenage drug use could warrant media attention, given how deep we were in our Monica Spring? These two adults were looking me straight in the face and asserting that somehow my past drug use could outscandalize President Clinton. What a joke. Once our meeting was over, I'd return to my office to field calls from journalists who spent their days neck-deep in President Clinton's sex life, trying to distinguish his lies from truths and determine whether or not he had pressured others to be dishonest. I'd take calls for an adviser who had to defend a boss who clearly sought intimacy outside of his marriage with female subordinates. Yet at that moment I was the problem.

My accusers told me to come back tomorrow and that I'd better think hard about my explanation.

I went back to Paul and told him what happened. He treated me with respect, expressing careful concern. I sat back at my desk, numb. But I sensed in a deep way that Paul had my back. I worked for a man who didn't panic or distance himself at the first sign of trouble, and right then, that felt like grace itself.

"YOU HAVE AN APPOINTMENT today for your FBI interview," Heather had told me that summer morning in 1993, once she picked me up from the West Lobby and we were alone in George's anteroom. She told me that the interview was pretty much *pro forma*, the last bit of the investigation process. Heather should know; she and every other new member of the president's staff who needed West Wing access had submitted themselves to FBI microscopes for their own clearances.

My interview couldn't come soon enough. I interned for Heather to relieve her of excess strain, but did so with a pink A (for "appointment") badge, the kind given to most interns that let them walk around the OEOB unescorted but not the West Wing. That meant I could never run even the simplest errands for her. There was no getting a breakfast bagel from the mess. No picking up journalists from Lower Press. No walking back visitors from the lobby or rushing through the downstairs rooms of the residence to drop off something in the East Wing. No luxuriating in a simple task that took you past the Oval Office and its chance for a glimpse at the president.

Talk about the ultimate backstage pass! It would be the most tender weight around my neck as it hung there and I entered the grounds, my access unfettered throughout the West and East Wings—well, at least through their hallways, for in a place like this, you just didn't walk into offices, or even their anterooms, unless you had a purpose, and even then you could feel trepidation as you bounced against the energies of people working with secrets as explosive as the biggest bombs a government, or a terrorist, knew how to build.

Back home, the FBI field agents completed their fieldwork, visiting my Troy neighbors, calling on my old high school principal. An FBI agent would then look me in my eyes and ask me questions about my past and discern for himself whether I should be entrusted with that kind of access.

"*Pro forma*," Heather said. "You're going to do fine."

And I did fine. I passed the background check, and I got my pass. For three and a half years, I worked for the Clinton presidency, either as a volunteer or as a campaign staffer. I never caused my bosses, or any other authority, to pull me aside and reprimand my conduct. I never took another recreational drug. Ever. They randomly drug-tested us, and I never worried because I never did drugs. Not after I entered college and started feeling that I owned my life fully and was in charge. Also, my mother, herself now a chemical dependency nurse, though never lecturing, managed to pass to me pamphlets and warnings about the health risks connected to drug use. The risks influenced my thinking more than any moral or legal arguments. The idea of my heart stopping from cocaine or of my developing depression or schizophrenia or having my genes affected by LSD or, worse, the idea of my mother standing above me as I fought for life in a hospital bed because of an overdose was a vision too painful to bear.

Five years later, they said I had lied about my drug use, and for the life of me I had no idea what they were talking about, no matter how many rooms I ransacked in my mind looking for clues. I burned inside, as confused and angry as a wrongly accused child about to be punished. How could my *I don't remember*s and *I don't know*s be sufficient to accusers who didn't know me? Who had no experience of my good faith?

"Admit to what you did, and you'll be okay." I knew this in my heart. Truth telling could bring temporary pain, but nothing like the walloping when you lie and hide and you are later found out. This is what they were trying to do with the president. The independent counsel asserted that he had lied in his Paula Jones deposition, that he had pushed others, including Monica Lewinsky, to lie as well. Obstructing justice, they charged. As they've said since the Nixon administration, it's not the crime but the cover-up that gets you into trouble. Not the act but the lies.

Though I believed in truth telling, I was starting to believe that in the eyes of authority, not submitting to authority was the worst crime of all.

SCHOOLS. WORKPLACES. References. I had listed them all so faithfully on the FBI forms. As if I were applying for a scholarship, but not, for with scholarships you had only bounty to gain, but with the investigation I had so much to lose: my prestigious new role as a White House volunteer who worked half days in the West Wing. But the process excited me: a background check! What a heady experience for a girl who was eighteen years old and so fresh to the ways of Washington. The U.S. government would spend serious money (a rumored $4,000) for FBI agents to interview the adults and peers of my past and ask them about my life to date, to find out that I indeed had a good reputation back home as a student leader.

I had dabbled in drugs, yes, but compared to my peers, so many of them Casey Jones wannabes with trouble ahead and trouble behind, I was virginal.

I don't have anything to hide!

But then a door creaked open. How could I have been so naive, so self-duped, to forget that deception had been so central to our teenage social lives that I forgot it was deception? I stepped out of the foyer of my mind and realized that I was wrong. Very wrong. I had a whole way of life to hide that had felt so accepted by my peers as to be admired that I had forgotten it was illegal.

When I was fifteen, I worked at a men's clothing store in Oakland Mall, a few highway exits away from my house in Troy. My manager, Kelly,* was a twenty-four-year-old woman I idolized: a postpunk, noir leather blonde with her hair tufted and soft, the dark roots showing. Kelly wore slick black eyeliner and kept her lips their natural pink. The ripped necks of her T-shirts revealed glimpses of her pale but voluptuous body. She seemed so mature to me, with her cool, appraising gaze of the world. When she wore ragged cutoff shorts with thigh-highs, so did I. When she said a band was cool, I bought the tape. I liked her so much and would have done anything for her approval.

She must have liked me too, because she took me dancing with her. I was only fifteen, but she took me to Detroit clubs and bars. She took me to the Shelter, the Warehouse, City Club, and other places whose names I never knew, with house music bumping and dark industrial raging and everything luminous with mystery. I drank and danced and thought I had sneaked into heaven.

But there was a problem—the clubs' age minimum was eighteen and over, sometimes twenty-one and over. Even though I looked older than my fifteen years, Kelly couldn't always smooth things over at the door. She proposed a solution, and I said yes in a heartbeat. Without thinking at all about the consequences.

Kelly drove me to the nearby secretary of state office, the one near my house, not hers. She gave me pieces of her mail and her Social Security card. Together we stood in line. She stayed with me all the way to the counter. Then came my turn. I stepped up, but Kelly didn't go far. I told the clerk that I had lost my license and showed her my proof.

---

*   This name has been changed.

There were no searching glances, no holding my papers to the light. They took my picture, then my money. The clerk gave me a blue temporary license and told me my permanent license would come in the mail.

This was 1990. This was precomputerized everything. Doesn't seem that long ago, but in terms of recordkeeping, we're talking ancient history. I left that office ecstatic that I'd found a fail-safe way to leapfrog the barriers that stood between what my friends and I were going to do one way or another: buy alcohol and go clubbing. More than that, I felt honored that this woman had done such a daring thing for me, this acolyte at her side.

Once the official license arrived, Kelly brought it to work and it was mine. Mine, all mine to buy booze anywhere, to walk up to any counter and whip out the card and let the transaction proceed. My parents never found out, perhaps because I never got into trouble. I knew better than to drive with it or to pass myself off as Kelly to an authority greater than a bar bouncer or a hotel desk clerk. This was for alcohol, clubgoing, and occasional hotel party room rentals only. Instead of driving into the city, we would pull into a liquor store off Crooks Road or walk into the Meijer liquor store off Coolidge—hoping not to see any adults we knew doing their grocery shopping—and buy what we wanted. When prom time came around, we needed to go only as far as the Drury Inn, half a mile from our high school, to afterparty in peace.

I felt so free. When you're underage, rules can feel as oppressive as the school cinder-block walls you can't scale. You don't think that one day you may be an enforcer, even a maker, of the rules. That even if you contemplate the spirit of the laws that constrict you, they still feel like constrictions, and any rule put upon you, as opposed to conceived or agreed to by you, can feel illegitimate. I look back, and I'm mortified that I committed this kind of fraud. Now I understand this as the crime that it was. Back then, I did not.

* * *

IN 1993, GARY ALDRICH WORKED in a cubby office on the fifth floor of the Old Executive Office Building. Until this appointment, I did not know there was a fifth floor to the OEOB. My usual elevators did not go there; his hallway felt tunneled through the eaves.

Aldrich met me at his door. Clean-shaven with tired eyes, he greeted me with a smile and a handshake. As he closed the door behind us, I realized his office was more cramped than our correspondence operation in 415. Natural light brightened the room, but not by much.

Aldrich asked me to sit. We quickly moved to the interview. Strangely, I did not worry about the fake ID, feeling confident that my friends wouldn't squeal on me and that none of the adults knew. Instead, I worried about my employment history: I had worked part-time jobs steadily since I was fourteen years old and two of my jobs ended in termination, both times because I wanted to take a day off and my bosses would not agree. Luckily, my forms listed jobs where I had been praised for job performance and had left on my own accord.

Aldrich spoke kindly to me. I started to relax. He asked his questions and I answered them one by one, seeing no sign of problems on his face or in his tone. There was something paternalistic about his way. When he broached the terminations, he nodded his head and smiled; his look let me know that getting fired from high school jobs for this sort of thing was no big deal.

Suddenly I sensed he was sympathetic, that he wasn't looking for reasons to stone me. The realization flooded me with relief.

Did he ask about drug use? Yes, he must have, but I could not remember what I said. *Come on, remember!* I could not. All I remembered were the work questions and Aldrich nodding his head, as in *of course you got fired a few times, who doesn't?*

He wrapped up the interview and led me to the door. It all felt good. Positive. As *pro forma* as Heather had promised. My God, I thought, this gatekeeper could have ended my White House dreams. Aldrich showed me to the elevator, and back into the world I went, never to return to that fifth floor again.

I still wondered about my fake ID. I had misplaced it the summer before college began. It had been an open secret at Troy High, like those of so many other students. But my peers tended to use hand-me-down IDs from older siblings and friends. No one else I knew had gone to the secretary of state and perpetuated fraud in so flagrant a way. If the FBI had discovered my mistake, it chose not to question me on it.

Soon my hard pass hung from my neck. I dived into my new life as an all-access intern, forgetting about the Aldrich meeting, the FBI process, and all the digging up and sharing that the questions had required.

I even forgot my first investigator's name until the FBI agent said to me: "You had Aldrich?" We exchanged looks. My first investigator had recently turned tell-all author. *Unlimited Access* was about what Aldrich had supposedly witnessed as a White House–assigned FBI agent—and it wasn't pretty. I had paid it no mind, assuming it was just more "cash for trash" that peddled rumors, distortions, and downright lies, the kind of book my bosses routinely derided. Maybe Aldrich's book revealed truths, too, but I walked past *Unlimited Access* in stores, having long since forgotten our brief connection.

Still, I had sympathy for the agent. When Aldrich entered the Academy, he could never have thought he'd end up in that fifth-floor cubby space, the kind of place that bred dust and resentment.

I resumed worrying about my own hide. In twenty-four hours, I would face my new gatekeepers, and I sensed no sympathy whatsoever.

* * *

NUMBNESS. In the hours after the counsel sit-down, Paul could have hacked off both my hands and I wouldn't have known. I answered the phones, played nice and attentive to all the callers, but on the inside, I was nothing but cold fog.

I clicked Drudge. I clicked through The Hotline, *National Journal*'s daily roundup of political news. Maybe I finished a letter or two for Paul's signature. One of my journalist friends called and immediately heard the change in my voice. I told him the story, and in a compassionate way, he told me to hang in there, that things would work themselves out.

But I cut the call off quickly. The numbness had settled in and I didn't want to talk. When the network newscasts started, I watched as Paul flipped between stations, gauging how much of our message had made it through their filters. I asked if he needed anything else. He said no, and I packed up to go home. He wished me well. I felt his concern. But I knew that he had much on his mind: the president's scandal thunderheads hadn't dissipated just because I was in trouble.

Back in my shared apartment, I lay in my bed with the bedroom door closed. My roommate was gone, and I huddled under my American flag print comforter, the one my mother had made for me herself before I left for England so I wouldn't be homesick. I pulled it up to my chin. I felt alone. No one could return to that counsel's office but me. No one could answer those questions but me. No one could truly share this burden.

I called my mother.

I had called her earlier, within minutes of returning from the counsel's office. She had listened. She had never interrupted. She didn't blame me for being stupid or for messing things up. I told her I had to go back to work, that I would call her later. I was in tears when she answered for the second time.

"Stacy," she said, "whatever happens, you're going to be okay."

"But, Mom—"

"They're not going to put you in jail," my mother continued. "Maybe you'll lose your job. But really, Stacy, is that the end of the world?"

She'd said the magic words. "No," I said, in a small voice. "It wouldn't be the end of the world."

For the first time, I believed it. I realized that as awful as this was, I would be okay. If they fired me, well, that would be shitty, but I would be okay. Fundamentally, at my core. They couldn't change that. Not when I knew my mother's love and support were unconditional.

"Do you even like that job anyway?" she asked.

"No!" I cried. I even laughed, with my mother connected to me like this, still keeping me safe somehow. Up until that point, I *had* thought the world was crashing. That the shame would be so crushing, I could never recover. I felt her heart as if it were my own, and never once did it stop doing what it always did: sustain me.

My mother told me she loved me. I told her I loved her, too. We ended the call. I lay in my bed and knew that I was going to live if I just kept telling the truth. And the truth was, I had been honest in that second background check. I presumed I had been honest in the first one. If only I could remember everything that had happened that first time. One's memory is so unreliable! The mind is a magnificent, never-ending mansion with so many wings closed off when not in use. I went at the locked doors with crowbars but still could not jimmy my way in.

I pulled my mother's comforter over my head tightly, so no heat could escape. Even though I felt her support, the fact remained that no one could face my questioners but me.

I TOOK MY SEAT on the counsel's couch. Time for my declarations of guilt and innocence. *Yes, I did drugs. These were the amounts. But no, I don't remember lying . . .* I refused to give in to their accusations, to just roll over without evidence, without my own brain telling me yes, what they said is correct. For the life of me, I *still* had no idea what they were talking about.

They deliberated.

"We're letting you off this time. But be very careful in the future," said the counsel. "Take this as a warning."

Then a blur as I walked across the glossy marble tiles, black, white, black, white, as passersby greeted me as if this were a normal office day. I remember the taste of cold stone building in my breath until I was several yards down the OEOB hall. I opened my office door, told Paul they were letting me off, and for one long extended moment we exhaled, both relieved.

We went back to work. He would never mention the incident again. Slowly, the adrenaline drained from my muscles and the tension subsided. But not before the most important lesson of my White House life crystallized: at some point, all of us would be asked to account for actions we've done in our lives and at a time not of our choosing—maybe at the worst time imaginable.

I thought about tax cheats, workers who take money under the table, drivers who ignore rules they don't like. All the ways we try to lie, to keep more than our share. As never before, I vowed to resist these urges. I vowed to myself that, to the best of my ability, I would never let what had just happened to me happen again.

AND THOSE ROOMS IN that massive rambling house of a memory? Slowly doors began to unlock. Years later, when I was far removed from the White House and Washington, D.C., I would walk down a leafy street and hear the creak of a door falling open.

Days after I met Heather Beckel and she said she wanted me to work for her and George that summer, I went to a party. A party with lots of young D.C. men and women, all working in the White House or on the Hill. I talked to a young man. I don't remember his name. Barely remember his face, even; we didn't become friends.

But we spoke about the FBI background checks, specifically the drug questions.

Just lie. That's what everyone does. Lie about it. Why ruin your chances of working at the White House over drug use? Everyone's done it. And everyone lies. Yeah, don't even think twice.

When I was eighteen, I thought this was good advice. I would lie. That's what I decided. I didn't want to blow the chance. What would everyone back home say? The possibilities loomed so large, and the fear of humiliation was so overpowering. At eighteen, I had core sets of beliefs, but I did not have the experience to know, with confidence, that I could tell the truth and truly be okay, no matter the outcome.

The memory of his advice did not reveal itself until years later. Not until I was out of the White House, out of Washington, and, quite frankly, far from the belly of the beast. Not until I felt safe. Not until I understood how fear motivates many of our worst mistakes. Back then, I didn't realize that my brain had sealed off that memory like a torpedoed section of a submarine. As if subconsciously, my system had determined that to do so was for my own good. That this was the only way to survive.

I think about this when I watch the president's denials of his relationship with Monica Lewinsky. We will never know fully the thoughts and decisions of President Clinton in those early weeks. But after my FBI experience, I have a heightened respect for the mysteries of the mind and how, under duress, we may not be as much in control as we think.

14

To Be Alone with a Powerful Man

To be young and female at the White House during that Monica spring. Imagine the questions each of us encountered from friends, family, and strangers: *Did you know Monica? Did the president ever come after you? Were you* really *a White House intern?*

Yes, and proudly. Instead of "presidential plaything," the title intern used to mean "student with a future," and many of us former students now served as special assistants or deputies to top staffers. We had names: Mary, Carolyn, Michelle. We had ambitions. Backstories. No two of us were alike.

But President Clinton had come to Washington with a reputation, and we knew it. Would he come on to us? To me, to you? I never spoke of this to any of my peers, but friends and strangers asked about "Bill" all the time. They made jokes about his sexual

appetites. I laughed the jokes off, knowing that they never represented the whole of my experience or how others saw me.

Then Lewinsky broke. The scandal placed a film between us
and your eyes. Suddenly we were all Monicas, our collective identity
absorbing her shame, the president's shame. Some of us looked at
the positive moments we had shared with the president with 20/20
hindsight; he made us wonder if he'd had only one thing in mind
when he paid us attention. We'd come in with our smarts and our
ideas, hoping to be valued for our professionalism; the idea that it
could all boil down to sex undercut so much belief in the president,
the man whose honor we defended daily, whether at work, or with
friends, or at the family dinner table. *Is he hitting on you, too? Does he
chase you around the desk?*

*No! And thanks for reducing me and my job to some awful cartoon!
We're not all like her!*

When a bad fate befalls someone we know, such as Monica
Lewinsky, we think of reasons why she was the victim and not we,
what we did right versus what she did wrong. I looked at Lewinsky's
situation and felt relief, knowing that I had never made the choices
she had. But in scandal-hot Washington, even the appearance of
impropriety could set off a feeding frenzy. I soon found out how
painful these sorts of accusations could be, not just to me but to
those who cared about my good name.

COME IN. SIT DOWN. *Here. At the head of the table. Videotape . . . on.
Audiotape . . . on. Now we administer the oath.*

The deposition began.

On March 18, 1998, I sat at a conference table in the Washington, D.C., offices of the advocacy group Judicial Watch, across
from its executive director, Larry Klayman, a pugnacious man
referred to in the *Washington Post* as a legal "gadfly" and known

around town for having sued his own mother. Judicial Watch was representing plaintiffs in a class action suit against the White House and the Federal Bureau of Investigation for violations of the Privacy Act, as hundreds of FBI files from previous Republican administrations had been wrongfully accessed by the Clinton White House—accusations now referred to as "Filegate."

I knew nothing about "Filegate" beyond what I had read in the papers, yet, there I was.

To my left sat the four attorneys representing the White House, who were there to protect me: Elizabeth Shapiro and Anne Weismann of the Justice Department, Sally Patricia Paxton of the White House Counsel's Office, and Marcie R. Ziegler of the firm Williams & Connolly, representing Mrs. Clinton, who was the named defendant in the case. Larry Klayman sat near a legal aide, Tom Fitten. Klayman would lead the deposition.

I wasn't nervous. I sat there with full confidence that my bosses and colleagues were working in good faith and if there had been any misconduct in the White House, it had been the work of an isolated, disappointing few, not the result of any culture of criminality. Never had I observed or sensed illegalities. The only time I had witnessed a boss play the slightest bit of hardball was on the movie screen, the scene in *The War Room* when George Stephanopoulos threatened the guy with the Bill Clinton illegitimate baby rumor that he would "never work in Democratic politics again" if he publicized the allegation right before the election. I knew my fellow staffers to be hardworking, professional men and women who took pride in winning by the rules.

Such as Paul Begala. He said he had no firsthand knowledge of the Filegate mess. Yet Klayman received judicial permission to depose Paul for his Filegate lawsuit. Why? Because Paul had made a joke about reading FBI files during a DNC event televised on C-SPAN. He made the joke, then turned to the camera and pretended to address a judge: "That was a joke, Your Honor." That was all it had taken to get Paul subpoenaed by Larry Klayman, and under oath. When Klayman asked him about our office filing system and Paul could not answer his questions satisfactorily, I, too, was subpoenaed.

Off I went to counsel Sally Paxton's office, which happened to be next door. In our attorney-client privileged meetings, Ms. Paxton walked me through the document search and prepared me for my time before the interrogator. I don't remember talking much; what was there to say? My office had few documents that satisfied the search criteria, almost all of them being photocopied news clippings—articles that happened to mention Filegate, and that was it. No smoking guns. I would share with Klayman how our two-person office worked, set up by a twenty-three-year-old (me) who had never set up another man's office before. Just a few file drawers with simple manila folders.

But office organization would account for only a fraction of Klayman's questions. Five minutes in, when Klayman asked about my temporary work experience in 1996, I knew how the rest of the six hours would go.

Q: Do you remember the names of some of the companies or individuals you worked for when you worked for Help, Unlimited?

A: Environmental Defense Fund.

Q: Was there a person there that you worked for?

A: I don't remember specifically their names. I worked primarily as a receptionist, and I don't remember their names.

Q: You're aware you're under oath, are you not?

A: Oh, yes.

Q: And being under oath means that you have to tell everything you remember?

A: Correct.

MS. SHAPIRO: Objection. She doesn't need to be reminded she's under oath. She took an oath when she began the deposition.

This would be the first of several times that Klayman reminded me that I was under oath, one of his bullying tactics that made me question his good faith and how committed he was to finding the truth versus advancing his own agenda. When Klayman asked me with whom I had discussed this deposition, I refused to give him my mother's last name.

Q: Before your deposition today, did you talk to anybody about it other than the attorneys sitting at this table?

A: Yes.

Q: Who?

A: I told my mother.

Q: Who is your mother?

A: Must I give you the name of my mother?

Q: Yes.

MS. SHAPIRO: I object to the relevancy of her giving her mother's name.

After four years of reading mail from the general public, I knew there were unstable and delusional folks out there, and if my transcript were to go public, I did not want my mother's full name easily tracked by someone with hard feelings. Of course, my mother had nothing to do with Clinton policies or proclivities, but those facts would not stop every aggrieved person with a will and a way. Klayman continued asking who else I had spoken with about the deposition, asking for their last names. Each time I refused. Ms. Shapiro, the lawyer of record, spoke up:

MS. SHAPIRO: This line of questioning is improper in that you haven't established what she might have talked about to these people. She's willing to tell you the substance of her conversations and generally who they were, her mother, her boyfriend, her father, but there's absolutely no purpose, other than an improper purpose, for her to be—naming people when you haven't established any relevancy or if you want to put the video and the transcript under seal.

MR. KLAYMAN: I'm not going to put the entire video and transcript under seal, but as a courtesy to her, if she didn't want to give us the names of her mother and father, I was willing to let her do that under seal.

I had asked a very simple question, whether she talked to them about this deposition. That is highly relevant.

MS. SHAPIRO: And she said yes.

MR. KLAYMAN: And I asked them for the names.

MS. SHAPIRO: That's right. And I think until you establish that there's anything relevant that she discussed with them, that their names are absolutely irrelevant and im-proper.

MR. KLAYMAN: Well, I may never know that if she's not being completely candid here today, and that's why you need to get the names of these people.

"I may never know that if she's not being completely candid," said Klayman, revealing the suspicion and the will to coerce that I was still not used to experiencing from perfect strangers. I tried not to let it bother me, despite the fact that we clashed like this over and over again for six hours. Back and forth we went, the process growing into stressful tedium as I tried hard to be accurate with my responses, scared to death of saying anything that could be construed as perjury. My answers frustrated Klayman:

Q: I'm not trying in any way to be critical here, but we can make this deposition, we can move it along faster if you try to answer my questions as best you can. I'm not going to hold you to exact precision?

MS. SHAPIRO: Objection. She's answered the best she can.

BY MR. KLAYMAN: Otherwise, I'll ask the questions sometimes three or four different ways. Sometimes doing that I've gotten an answer. Now, that can make this a very long process?

A: I understand.

Q: And it will be my position, Ms. Parker, that if it becomes a long process that we will ask the Court for permission to bring you back again. So I ask you if you will move this along as quickly as possible?

MS. SHAPIRO: Objection to trying to intimidate the witness. She's answering fully.

MR. KLAYMAN: I'm not trying to intimidate her. I'm trying to get this thing to move along.

MS. SHAPIRO: If you can ask more precise questions, maybe it will move along faster or maybe relevant questions.

MR. KLAYMAN: Certify this, improper remark.

Certify this. You're on notice. You know you're under oath. You can respond. Certify this . . . Later I asked the attorneys what "certify" meant in this context and they laughed it off, having no idea what Klayman had been talking about. *Has anyone ever instructed you* [*to say what you just said*]? . . . *Were you programmed to make that response?*

Aaaagh! If only Klayman could understand that I was told to tell the truth! Sally Paxton had never rehearsed me. Nobody had rehearsed me. They had oriented me, told me what to expect. But as far as instructions? They had simply said, "Tell the truth."

And the truth was, I felt blameless. I felt bolstered by support—from Paul, from the lawyers—because it wasn't my fault. I was simply caught between two competing realities. Klayman spent a good deal of time asking me about White House opposition research efforts, acting as if we were all players in a spy thriller—as if every time we walked down a hallway or read through a file, you'd hear a snake rattle hiss. Our research guys helped us defend ourselves when the Republicans piled on, for the researchers found previous "quotes and votes" that revealed the hypocrisy behind many of our detractors' criticisms. But we did this all aboveboard, using public records; I was never aware of any illegal intelligence work, such as the combing of FBI files.

Yet Klayman seemed to think the White House was filled with criminals digging for dirt. Circumstances that I would write off as happenstance—such as his later grilling of a research staffer, Tom Janenda, about the presence of an aide of the First Lady in his office suite—Klayman seemed to suspect had the most sinister meanings and motives possible. At the time, I was unaware that the president's personal, non–White House lawyers had retained private detectives, such as Terry Lenzner, who had apparently looked into the personal lives of women who had accused the president of adultery. Or that some Clinton defenders let it be known that they would go scorched earth if Congress pushed for impeachment, revealing affairs and sexual proclivities that would humiliate the president's accusers. But I was totally unaware of this dank level of intrigue. I only knew what I experienced in the White House, and Klayman's version of reality seemed paranoid and delusional.

Eventually Klayman moved on—but to arguably more shady territory, as he asked me about Vince Foster and Commerce Secretary Ron Brown, whose deaths had became fodder for Clinton-fearing conspiracy theorists. He even asked me about the 1997 Starbucks triple murder, another tragedy conspiracy theorists attempted to link to the Clinton administration. These questions yielded little until he asked about the one man that anyone with ambition, achievement, or luck in Washington, D.C., knew: Vernon Jordan.

Yes, I knew Vernon Jordan.

Klayman asked me how, leaning in, as if he had struck oil. Given what we now know about Jordan and the president and Monica Lewinsky, I could see how someone might wonder if Jordan had tampered with me. But this was impossible, for the last time I had spoken with him was before Monica hit, and never once had we spoken about Paula Jones or her suit. And I sat there angry inside, as Klayman forced me to answer questions under oath that had nothing to do with the stated reason for my deposition.

The attorneys tried to shut down this line of questioning quickly. We took a break, but as soon as the videographer hit "Record" once more, the questions about Jordan continued, as Klayman tried to find out if he had tried to influence my testimony:

MS. SHAPIRO: Can you, Mr. Klayman make a proffer on relevancy of this question?

MR. KLAYMAN: The proffer of relevancy is that there has been reported if not demonstrated conduct that Vernon Jordan has allegedly engaged in witness tampering. That's the relevancy among other areas, but that's as far as I'm going to go, but that's enough certainly to allow me to ask these questions.

MS. SHAPIRO: And how does that relate to this matter?

MR. KLAYMAN: I think you know, Ms. Shapiro, you're a pretty bright woman.

BY MR. KLAYMAN: Now, Ms. Parker?

A: Uh-huh.

MR. KLAYMAN: Can you read back the last question before we convened this deposition?

(The reporter read the record as requested.)

THE WITNESS: Either once or twice, we've gone out to lunch, and there were two or three times we've gone out to dinner.

Q: Was anybody with you when you went out to dinner with him?

A: No.

Q: Was anybody with you when you went out to lunch?

A: No.

Q: During the times that you went out to dinner, when was that?

A: I don't remember exactly when, but it's been over—over the last three to four years. The last time—I'm trying to think of when the last time was. I haven't seen him—the last time I remember seeing him was during the inaugural last year.

Q: Have you talked to him since then?

A: A few times, yes.

Q: When did you talk to him?

A: The last time I spoke with him was last year at the—I had already started in the White House. I don't remember exactly when.

Q: At the time that you went out to dinner with him and went out to lunch with him, were you in any way asked to be a witness in any legal proceeding?

A: No.

Q: Did he ever offer to get you a job?

MS. SHAPIRO: Objection. She's testified as to the things that would matter to your proffer or relevancy.

MR. KLAYMAN: I don't want to give her the answers to all my questions.

BY MR. KLAYMAN: Did he ever offer to get you a job?

A: He's never offered to get me a job.

Q: Has he ever assisted you in getting a job in any way, however small?

A: I remember he kind of tried to help me figure out what I'd be interested in doing. But he—I'm not aware of any phone calls he's ever placed on my behalf or any actions he's done on my behalf to assist me in getting a job.

Eventually, Klayman moved off Vernon Jordan. He did not take long to ask me about President Clinton himself:

Q: Have you ever met with the President one on one, you and the President?

A: What do you mean by that?

Q: Where nobody else was present?

A: I mean I've never scheduled a meeting with the President. I mean, there have been times when he's dropped in to—when I worked in George's office, he would occasionally walk through looking for George. But I don't know what he was doing, but I assume he was looking for George, and those are the only times where I've been in that room—that in that room, he and I were alone.

The attorneys cut that off quickly, but I had nothing to hide. My inner Dirty Harry emerged, and I wanted to say: Go ahead, ask me about President Clinton. You don't think every journalist in D.C. hasn't sniffed up every White House woman's skirt looking for another Monica? What makes you think you would be the one who finds another poor girl to humiliate before the world? Because that's how a woman would feel the moment she became like *her*, like *that woman, Miss Lewinsky*.

Sorry, Mr. Klayman, I thought. I'm not her.

AFTER THE DEPOSITION, the lawyers told me I had done a good job. One asked me if I considered becoming a lawyer myself. I smiled and said maybe. The lawyers told me not to worry too much about the ground that had been covered. But there was still something pinched in their expressions. An understanding they must have had that this thing was not over. Neither for them nor for me.

* * *

I DON'T KNOW WHERE I saw the March 27 *Washington Times* first. But I saw the headline, and my picture, staring back at me from the vending box on Seventeenth Street somewhere in front of the red-brick anonymity of the New Executive Office Building. We received the clips and hard copies of the major dailies at work, so I didn't keep subscriptions and I rarely bought from the street.

But every morning I checked the boxes. I wanted to know which headlines, which pictures would be absorbed by passersby regardless of purchase. Like Paul, I gauged daily how much of our message made it to that most influential of news real estate, the top fold, flush right.

I saw my picture on the top fold of the *Washington Times*. Not flush right, and not a screaming banner headline. But above the fold. I found consolation in the fact that this was not the *Washington Post*, the paper of record, but the slanted and sometimes sensational *Times*, with its open appeals to a conservative readership, the one in the red box that I never slipped quarters inside. Klayman had released my full transcript to it, and this was what it had found newsworthy:

JORDAN AIDED HER, TOO, SECOND EX-INTERN SAYS:
TESTIMONY SUPPORTS LAWYER'S ASSERTION
By Paul Bedard
A former colleague of Monica Lewinsky's as a White House intern testified that Vernon E. Jordan Jr., presidential confi-dant and Washington superlawyer, gave her job counseling over a series of intimate dinners, The Washington Times has learned . . .

Intimate dinners? I knew what I looked like there, regardless of whether wrongdoing of any nature could be proven.

"I've seen him," testified Miss Parker in a deposition taken by Judicial Watch Inc., which is investigating whether the White House looked at secret FBI files to obtain damaging information about Republican critics. "I've—he's offered me counsel in the past. I have, I have, I consider him a friend."

I imagined what readers must have assumed: *Why is she stammering if nothing is wrong? And a "friend," eh? What kind of friend? Something must have happened.* If they were at a loss to think of what that "something" might be, journalist Bedard added:

Mr. Clinton and Mr. Jordan, who often play golf and spend time together, are known to swap randy locker-room tales about women. "We talk like men," Mr. Jordan told an interviewer in 1996. "There's nothing wrong with a little locker-room talk."

For example, Washingtonian magazine reported that in 1995, when Mr. Clinton was seated next to "an especially attractive blonde" at a White House state dinner, he "laughingly ordered Jordan to keep his hands off the woman because 'I saw her first, Vernon.'"

I saw a thousand smirks as readers enjoyed the jab at Vernon Jordan, caught consorting with a lesser Monica. "Lesser" because Jordan had less power than the president. Or "lesser" because they were beholding what must have been the worst picture to ever exist of me. Ever. The *Washington Times* had printed a video still from the interview, effectively bloating my face so much that I looked thirty pounds heavier. Worse, I had decided not to get my hair done, and my kinky curls looked limp and fuzzy. I had made the "no hairdresser" decision in defiance, thinking that I wouldn't spend the money and make the effort just because Larry Klayman had summoned me to his quarters.

He showed me.

But here is where I lucked out. Way down the article, way after the jump, in the second-to-last paragraph, colleagues stuck up for me:

> Miss Parker graduated from George Washington University with honors, and is described by others at the White House as well liked and respected.

For fifteen paragraphs, I'm just another girl used by men. But if the reader could just hang on until graf 16, another story line was revealed, such as the possibility that I might have been worthy of mentorship and that there was honor there to protect.

MARK HALPERIN SAID TO ME once, when talking about sex scandals, that no one ever knows what happens between two people in a bedroom unless they're under the bed and even then they don't know, not fully. I was never alone with my bosses in bedrooms, but I was alone with them in offices, in living rooms, in Washington, and on the road. In these rooms, I'd sit and talk, laugh, listen. Sometimes I'd feel the male boss go over the line, but if I did, I found a way to end the meeting as quickly as possible. I never felt coerced. I always felt free to choose my next move. Always. And for each time that such visits resulted in a warm, mentoring conversation, I was willing to risk the few times I felt like the lead in *Escape from the Lion's Den*.

Why did I enter these situations? Because I was asked. I was flattered. I was so eager to learn and be included that I was thrilled that these important men would make time to visit with me. Even if their intent was sexual, I never wanted to believe this was fully the case.

Especially with powerful men. Like a stack of TVs turned to different stations, they could exude nurturing love as much as they did sexual desire, and there were times I chose to keep my eyes on the nurturing love show and ignore the others.

* * *

FROM THE *NEW YORK POST*, Page Six, April 22, 1998:

> President Clinton wasn't the only middle-aged man who enjoyed the company of young interns like Monica Lewinsky. His golfing buddy, D.C. Superlawyer Vernon Jordan, had a favorite intern too. Her name is Stacy Parker. Parker, 23, was forced to reveal some of the details of her "business and personal" relationship with Jordan last month when she was questioned under oath in the Filegate lawsuit. . . . When PAGE SIX reached Parker yesterday at her White House office and asked her about her relationship with Jordan, she replied, "Unfortunately, I'm not really allowed to talk about it."

The e-mails poured in. "Look, you're a star," wrote Jeff Eller, a former White House staffer and friend of Paul, bucking me up, letting me know this was okay. "You made the Hotline!" fawned Ann Marie, my friend from college, who, like so many of us, knew that publicity was attention, a power of its own, even if the words were the opposite of what any young female staffer wanted written about her. White House counsel Lanny Breuer pulled me aside and in his friendly way reassured me, told me that one day I would look back at this and laugh, that for all time I had been gifted with a great cocktail party story—my small, mildly distasteful role in what the journalist and presidential counselor Sidney Blumenthal dubbed the "Clinton Wars." I got off lucky, he said, and I should do the smart thing and forget about it. Keep talking about it, and you kept the story alive.

LUCK KISSED ME HARD. My reputation was good enough to kill the story. If I had been seen as loose somehow, journalists would have kept sniffing, looking for gold with a second intern scandal. But the story dropped.

Almost.

"Carol, hi! Yes, it's Wanda! Oh, my gosh, do you listen to Don Imus? I was driving to work, and this morning he mentioned Stacy! He was kind of an asshole about it, but that's how he is, an old grouch, you know? He was making jokes about her and Vernon Jordan. I don't remember exactly . . . we were just so shocked! Is she okay? Well, not everyone gets to be made fun of by Imus on the radio. Kind of a badge of honor."

My mother's sister lived in Las Vegas. I had only just met Wanda at the Kansas family reunion, and suddenly I was a ghostly presence in her car.

My mother called me immediately. So did Nana. They did not blame me; they blamed the men. Nana—more than a decade older than Jordan—kept her words cool, even cold, when we spoke about this. *What's that old lion doing around my granddaughter?* She immediately keyed into his motives in a way no one else around me had done, at least not to my face. She didn't upbraid or belittle me. But I could tell she felt the responsible one was Jordan, the older man who should have known better.

I kept these conversations short, in no mood for postmortems. I could do this with my family. Unfortunately, this was not the case with my boyfriend's mother.

Jonathan called me shortly after the story ran in the *Washington Times* to announce that a shorter version of the piece had run in London's *Daily Telegraph*. I laughed, quietly dazzled that my bit role in our national soap opera would get international play, wondering if any of my Oxford friends would come across it, learning exactly what I was up to these days. What should have been occasion to laugh became somber when Jonathan told me the next part.

"Stacy," he said, "my mother read it."

I felt both hollow and heavy, feeling the weight of his mother's deep disapproval, the judgment of a woman I felt had only recently embraced me.

"She thinks you should write a letter to the editor and consider suing for libel," said Jonathan.

"But the lawyers told me I can't say anything else right now," I pleaded. "If I do, I'll make it worse. You *know* I want to defend myself—"

"She wants to talk with you," he said.

"Right now?" I asked.

"Yes," he said. "She's right here."

I felt panicky. She and I *never* talked on the phone. I felt nervous whenever we talked in person—she had a kind manner, but there was steel in that softness, and she could speak at length on a passionate subject—the last time having held forth on the evils of abortion—without stopping for breath or for more than nods of agreement from me. This woman, so slim and petite, with her hair always wrapped in a beautiful silk scarf, always attended to the needs of her five sons, Jonathan being the youngest. She exercised authority without ever raising her voice. She was a force. I feared speaking to her, unable to do what I knew she would expect. Paul was in a meeting but was due back any minute. This was not a good time.

"Hello, Stacy, how are you?" she said with a warm concern that I wanted to respond to, but she continued, "I read the article, Stacy, and I'm afraid you must seek redress. All one has is his reputation, and they have tarnished yours with these terrible rumors and innuendo."

I felt the shame settle in hot. She was speaking the truth, but what was I to do? Lanny Breuer's words were in my ears, and I was sitting in a beehive of staffers focused on survival—this was not the time to make righteous stands about personal honor. But she continued, "You must consider writing letters to editors of these papers, Stacy. You cannot let this go unanswered."

I started to explain to her the legal strategy at work, but she was having none of it. "Listen to me, Stacy: you must avoid these situations in the first place. Jonathan tells me that it is indeed true that you dined with Vernon Jordan alone. There is no proper reason for a young lady to be alone with a man so clearly her senior, who has all the power, you see? It is not fair, of course. But you mustn't be alone with these men," his mother said.

"What if I need to get a recommendation?" I asked.

"Stacy, you cannot tell me a reason that is good enough. Nothing is worth your reputation."

I placated her with "Yes, yes, you're right," feeling numb and caught between her push to save me and my push to save myself because still, I thought, she could not be right. Not fully. At least not for a girl like me who wanted to change the world. Maybe for virginal debutantes who would marry well and live safe, correct lives. But was I that girl? How could I learn how the gears worked if I never got close?

Jonathan returned to the phone. I pleaded my case that as long as I worked at the White House, I had to stay quiet about this. To please explain this to his mother and promise her that I would try to do better.

I thought of the vacation that I had taken to Israel with Jonathan, and how we bought his mother a gorgeous gilt icon of the Virgin Mary that she hung on the wall. I thought of Our Lady beloved by billions, but in that picture she seemed so trapped in two dimensions, only a slice of her true self. To be worthy of Jonathan, was I supposed to flatten myself, too?

I felt sick inside. I agreed that I needed to defend myself; but to live as she suggested, well, I might as well stay locked up in a pretty room. Good advice for the girl who wanted to avoid blame. The problem came when you wanted both knowledge and blamelessness. I realized that one would have to be sacrificed for the other.

* * *

"Bimbo eruptions"—that's what Betsey Wright, the longtime staff mother to Bill Clinton, called the phenomenon of hairsprayed women standing before clustered microphones and announcing to the lights and the black glass eyes that *yes, the president came on to me.*

An eruption: creating the kind of mess staffers had to wipe down with bleach. That I understood. But to use "bimbo" like that, to define the accuser as loose, as less than, resorting to the ugly way that women have been silenced since the dawn of time—I found that problematic to say the least. If the president was so innocent, why smear the accuser?

Paula Jones said that in 1991, the governor of Arkansas had unzipped his pants in a Little Rock hotel suite and asked her to *kiss it.* She said she could identify a physical characteristic unique to the president's groin to prove that her story was true (enough). That's what she said. And yes, it was a "he said, she said" situation, a frame like "civil war" to warn us outsiders to give up, that there was no hope here in distinguishing the aggressor from the victim, if such finite categories ever truly exist. A good strategy for anyone who hoped to blur the line that separated his will from hers, force from fellatio.

I watched Paula's press conference and took the cue from her look, from the snarky commentators afterward that she was "trailer trash." *She is not us*, we reassured ourselves. She said she wanted an apology, but she really wanted money and was willing to be used by the president's enemies to get it. I hated the nasty characterization and, in it, the contempt for her origins, but I watched Paula speak and thought, Why is she trying to take our president down? Lose at the ballot box, so you turn to the courts: that's the modus operandi of our enemies. That's what made it so easy to ignore Paula. I did not take seriously the threat of harassment, for no one had pushed himself on me in the White House proper, no one had created a situation I could not escape from with a smile.

What motivated Paula Jones to pursue her civil case, I don't know. Was she wrong to do it? I don't know this either. Should there be no price for a man lording his incredible power over a woman, even if he had no intention of using it? Power that can feel to us like nuclear weaponry?

In school, we learned about "mutually assured destruction"— the understanding American and Soviet leaders had that the other side would be loath to use its nuclear missiles, given that the response could equal its death. That was what we reasoned, of course. That was something we never knew for sure. The fact remains that government to government, man to man, man to woman, woman to woman, we are still Kremlinologists of each other, poring over the photos, the clues, looking for the real truth of situations— truth that never comes out of our mouths perfect and whole. Even those who spend their whole careers decoding the other still make mistakes. We miss major developments in each other's countries, the pending failure of entire systems. No matter how "expert" we become, we still get it wrong.

There's only so much we can read of the other.

15

Bunk Beds in the Hamptons

ANOTHER LATE AFTERNOON IN OUR OEOB OFFICE AS PAUL REACHED for the remote control and unmuted the TV for the independent counsel update. Work stopped as we watched the latest CNN report.

There was Kenneth Starr, dragging Larry Cockell of the Secret Service before the grand jury. A seventeen-year veteran, Larry Cockell served as Clinton's special agent in charge (SAIC). He was the tall, distinguished-looking black man next to the president in close shots of him exiting his limo or working a crowd. The SAIC stayed mute as he worked, but his eyes saw all as he moved and monitored the fluid, potentially wild flow surrounding the president's person. Larry Cockell must have had a million picture stills in his memory of such moments, public and private, and just as many snatches of overheard conversation. The independent counsel knew this and summoned him to testify.

I hadn't met Larry Cockell and had seen him only in the hallways and at events. But I admired him from afar—as I imagined much of America did—watching him enter and exit the federal courthouse as Clint Eastwood's character in *In the Line of Fire* came to life on the evening news. The fire this time was legal jeopardy—and the future of Cockell's previously untarnished career in law enforcement.

My heart hurt for him. Cockell had signed on to protect, to take a bullet if need be. He didn't sign on for this. The law was asking the law to violate codes of confidentiality, of trust, just so an investigator could know the secrets of another man's sex life. The investigators had to be embarrassed inside. I understood the desire to determine whether or not the president had perjured himself or obstructed justice in the Paula Jones sexual harassment case, but the independent counsel's hunger to damage and embarrass this president seemed insatiable.

Paul muted the TV again. We went back to work.

Later that day, the phone rang, and on the display was Brian Alcorn's name and extension. A spark went through me. Magic happened when the Advance Office called.

"Would you do the RON in East Hampton, New York?" asked Brian.

I could have kissed him through the phone. Six months into the Starr/Lewinsky disaster, and I leaped at the chance to leave the bunker.

"The good news," Brian said, "they're RONing with the Spielbergs."

Steven Spielberg! I was already leaving the White House early one day a week, at 6:00 p.m., to head to a Bethesda fiction workshop. I already knew that what I really wanted to do was write. And now, a writer's dream: face time with a master storyteller.

"Bad news is that this will be a split RON. We don't have enough rooms at the compound. We're renting hotel rooms and trying to rent or borrow additional homes. This is going to be spread out, sorry about that."

Brian didn't ask me if I was up for it. I was supposed to be. That's what advance people did: they worked things out. Yes, I said. Yes, without a doubt.

DEPENDING ON WHOM YOU ASKED, RON stood for "Remain Over Night" or "Rest Over Night," referring to both the location where the president spent the night and the advance people tasked to look after the site. On the daily presidential schedule, the last entry was the RON. If the president was at home, the schedule read:

> RON   THE WHITE HOUSE
>   *Washington, DC*

If he was on the road, the schedule might read, for instance:

> RON   THE FAIRMONT HOTEL
>   *San Francisco, CA*

As the RON advance person, I never made decisions out of thin air. Protocol dictated who would sleep closest to the suite and who received the best rooms. The problem was that hotels configured floors and rooms differently, as did private hosts, calling for independent judgment on the ground. My bosses reviewed and approved my decisions. But I was the one who saw the rooms. They relied on my judgment. Unless I made obvious mistakes, what I said went.

After Brian slated me, Lito quickly called. He was one of the dapper Filipino valets who assisted President Clinton. Lito would work this trip, arriving just before the traveling party. I wished I could have impressed him with my Tagalog, my stepfather's native language, but I knew too little to even try. Lito knew I'd speak to the Spielberg compound staff first, so he gave me his list:

1. Sodas, water.
2. Lots of fruit, esp. baskets; Diet Coke; diet ginger ale.
3. Fresh orange juice; fresh grapefruit.
4. Flowers, but no pollen, no smell (i.e., carnations, roses).
5. Pillow: foam, not feather.
6. HRC: separate bathroom, separate wardrobes.
7. Is there a chef? Suggest:

-canned tuna	*-melons*
-chicken	*-berries*
-salad	*(strawberries, raspberries)*
-soups	*-banana*

8. Room temperature—know how to control.
9. Who is cleaning the house after they leave?
10. Laundry facilities?
11. Set up bar w/assorted (vodka, gin).
12. Iron/steamer.
13. Food for Chelsea:

-mac & cheese	*-salads*
-egg white	*-assorted cheeses*
scrambled eggs;	*-vegetable plates*
omelets	*(broccoli, celery)*
-skim milk	*-soups*

Back to convincing myself that the tension in my stomach was more excitement than dread. I was still new to advance. There was so much to learn, so quickly, with no time to apprentice. Swim, girl, swim, they said. Or just don't mess up.

* * *

THE ADVANCE TEAM flew into Islip, Long Island. We rented vehicles at the airport, prearranged and prepaid by the Advance Office, and drove the rest of the way to the Hamptons. Although it was a weekday in July, the route was busy and alive. The two-lane road filled up quickly.

Travel through each "Hampton," and it's easy to see why people pay millions to live there. The green of the tree canopies. Blue sky that stretches up and over the ocean. The more we drove down the lanes with the tall hedges and the seaside so near, the more I forgot Washington and its peculiar stresses.

The advance team would stay in a small Amagansett inn. Each room opened out onto the pathway hugged in hard by overgrown shrubbery. Outside, I worried that bugs would jump into my hair, unbeknownst to me. Inside my room, ocean air and tramped-in sand permeated all smell and taste, as did the rot that the sunshine through windows would never kill. They were not typical POTUS advance accommodations, but that was because POTUS was not staying there. I unpacked my suits and dresses and joined the others outside. Off to the Spielberg compound for our first walk-through.

Mitchell Schwartz was our lead, and he got behind the wheel of the staff minivan as the rest of us piled in. Mitchell was smart, funny, and good-looking, with a laid-back persona. I knew of him because he had dated Heather's good friend Wendy Smith, who had dated George before he had left the White House. As he drove, Mitchell talked about restaurants and bars, possible "OTRs"—off the record sites that the first couple might go to spontaneously— that we'd want to check out first, some of which I knew from gossip columns. I listened as I took in the scenes of exclusivity. I imagined Manhattanites behind the tall green leafy hedges, luxuriating in all their space and privacy, the greenness denied them in the city.

I hoped that Mr. Spielberg would be home. My fantasy was

silly, but I didn't care: maybe the Great Director would get one look at me and realize that it was *me* that he had been missing his whole life. He would offer me a job on the spot, or a screenwriting tutorial, or just a chunk of cash just to go off and write my story.

There was nothing wrong with dreaming, but I'd felt myself crossing the line between dreams and delusions too much that spring. I went out to lunch with CNN's Ed Turner, the creator of *Talk Back Live*, and I thought, maybe he'll give me a show. (Though I had zero media experience.) I took a Shell executive on a West Wing tour, and I thought, maybe he'll create a high-paying job for me in London (when the man had just met me). Any time I did something nice for a VIP, I wondered what he could do for me to get me out of my job, as if all these powerful men carried the pixie dust to make our dreams come true.

Ah, power. We thought it promised so much. I knew this because I worked in the White House and I saw the way people were around the president. They rubbed their hands together. They schemed. They thought he could power any machine.

We pulled into the compound gate. Mitchell jumped out. He spoke to the security man for a moment, and the gate opened.

The Secret Service agents already had the run of the place. They buzzed around, looking important, being important. Same with the White House communications technicians (WHCA) setting up shop.

We parked. Stepping out of the van, I took in the grounds. Four houses and outbuildings surrounded a long circular drive, the grassy innards as big as a football field. Ocean air blew through. Blue sky above, birds flying from the sea just over the dunes. This was my first time on a compound. Pretty cool.

Spielberg's staffers met us—his government affairs guy and the manager of the grounds. I overheard someone say that the Spiel-bergs were not there and hoped no one caught the disappointment on my face. We would, however, go inside three of the houses: the designated Clinton guesthouse, the staff guesthouse, and the Spiel-bergs' home. I tried not to look impressed.

The first full walk-through began. We walked in a gaggle, at a deadly, slow-motion pace. This was before BlackBerrys, so my teammates who had nothing to do with the RON had to trudge along and fake interest as best they could. Often only the leads talked during full walk-throughs. They tried to make as many mutually agreed-upon decisions as possible while everyone was in one space together, looking at the site. When we approached the houses, I came up to the front. I was introduced, but Mitchell did the talking, and that was fine, because he seemed to trust me, and I didn't feel that I had to make a show of proving my compe-tence.

The Spielbergs' home interior looked like a Colorado ski lodge, with high vaulted ceilings and exposed beams, the second-floor rooms leading to a balcony that looked down on the living room. Though the home was furnished with an eye to rustic comfort, a few framed photos of his beautiful kids were the only clues that this house belonged to someone and was not a showcase model. We did not stay long.

We moved on to the POTUS guesthouse. Not much different from the Spielberg house. Since I was the RON, I could go up with the compound supervisor and peek around the bedrooms. *Two bathrooms, check. Adequate light in the FLOTUS bathroom, check. Room for FLOTUS wardrobe, check. May we have hypoallergenic pillows for them? Check. And also, no aromatic flowers in the arrangements, due to allergies. Check.* I promised to provide a written memo of all of our requests, including Lito's food lists, the next day. So far, everything looked to be in order.

We talked of bed space for the critical traveling staff. The an-swer was that yes, there were plenty of beds on this compound. But were there enough for us? Mitchell assured the host staff that we would make whatever they had work.

I matched up the manifest with the list of beds and rooms available in the second compound guesthouse—the one designated for POTUS staff accommodation. Certain aides had to sleep there; they couldn't be housed in nearby homes or motel rooms. I had a problem. In the staff guesthouse I had two master suites and two kids' rooms, each with two sets of bunk beds. Four, possibly five high-ranking staffers of the American president would have to sleep barracks-style. These body staffers sometimes got stuck with unfortunate sleeping arrangements. They knew the deal, but there was the real fact that these doctors, agents, and valets were on the road constantly, and after working grueling shifts, I knew they wanted to collapse in comfortable rooms of their own, with doors that closed them off from everyone else, if only for a few hours. I looked at the rooms with bunk beds and shook my head. Who would want to sleep like that? Especially when you're sleeping in a billionaire's luxury compound.

We stood around the staff guesthouse, broken up into small groups. Mitchell was off frying bigger fish. The site people looked bored, wondering when they'd get to leave and check out their event sites. I kept sketching a house diagram on which I'd map out room assignments, then show my bosses for approval. Once it was finalized, Brian Alcorn would include the diagram in the official trip book. I tried to get the sketch done before I was called to leave for the day.

The military aide (MIL AIDE), looking fresh, scrubbed, and affable, approached me with a big smile. The MIL AIDE carried the "football"—the attaché case of nuclear codes and other crucial information. He also served as the president's military staff liaison. He would be on duty during this trip, so he came early to walk the routes and familiarize himself with the sites. The MIL AIDES were friendly yet mysterious men to me; even as a staffer, I didn't know the breadth of their responsibilities. But right then, the MIL AIDE was all "How you doing, and haven't we worked together before?" No, I said. I'd seen him around, but never on a trip.

"May I have a master bedroom?" he asked, all blue-eyed twinkle twinkle. "You know, my wife is coming up. This is our anniversary. It would be really great to not have to share."

"No problem," I said. Why shouldn't the man who carried the football have a master bedroom, especially with his wife coming? I penciled his name into one of the master bedroom boxes on my diagram.

The doctor approached me next. He was advancing this trip for the president's chief physician and for his own time on the ground.

"Stacy?" he asked, somehow knowing my name already. "I see there's only two master bedrooms. You know, Dr. Connie will be here."

Say no more. I'd only done four RONs prior, but I knew all the medical staff were anxious to keep her comfortable. A diminutive Filipina, but with a commanding presence, she made her needs clear. Even if he hadn't smiled sweetly at me, I would have given her a master bedroom: she was the lead physician, and she was the lone female in the tight package.

Only the kids' rooms were left. According to the manifest ranking, I had already accommodated the two highest-ranking staffers. Who was left? The valet. Three Secret Service agents, two of them supervisors—the SAIC, Larry Cockell, and the deputy SAIC, Donnie Flynn. Kris Engskov, the body aide. I assigned the agents one bunk-bed room and the valet and Engskov the other room, but this decision called for backup. Maybe they'd want me to search for hotel space. Maybe they'd want me to try to scare up additional beds on the compound.

I consulted with Brian, my advance desk, then Kirk Hanlin, the trip director. They told me that Kris Engskov would stay in an off-site house, but other than that, my mock-up stayed as it was. I finalized my room list, finished the diagram, and faxed it all back to the White House for inclusion in the trip book.

* * *

THE COMPOUND PREPARED. There were agents and WHCA technicians everywhere. Men and women in khaki shorts walking back and forth. They had so much work to do to make this site ready for POTUS and FLOTUS. I was privy to few of their protocols.

I stood outside the Spielberg residence in a shadeless spot, waiting for the assistant to emerge. I had my memorandum of the first family's requests. I had every reason to be there, but I felt weary of trespassing. I bumped against invisible fences of protocol all the time, the fear that I was in the wrong place at the wrong time. The feeling that I might be staff, but I was only staff.

A golf cart approached. My Secret Service counterpart sat inside, next to one of the Technical Security Division (TSD) guys and another agent I didn't know. I pushed my sunglasses up past my forehead. Wait—the man driving was Mr. Spielberg! He was talking, smiling, and the agents were laughing. Mr. Spielberg pointed something out. The agents nodded.

As they drove closer, I smiled, but I didn't wave. Mr. Spielberg didn't know me. I couldn't risk looking overeager.

They drove closer. I saw Mr. Spielberg's face, and I saw total delight.

They approached. Close. Closer. There they were.

They passed me. They didn't even slow down. I got a quick nod from my counterpart, but that was it. They were being driven by Mr. Spielberg, and that's where their attention would stay.

I watched the golf cart ride the turn. The men held on, laughing, having the time of their lives. I put my sunglasses back down. I was literally in their dust.

Mr. Spielberg knew what I knew: agents were special. They had mystique. They had skill sets we didn't. They were also in the process of force-fielding the man's property, using state-of-the-art toys to do it. Of course he wanted to hang out with them. At least to know what the occupation force was up to.

I left work early every Wednesday evening to go to my fiction workshop. I had asked Paul if I could do so, and Paul, in his kindness, had said yes. My dreams of influence shifted from the political theater to theater itself. My bosses were in the storytelling business, too. But I wanted to be free to tell my own, as I saw fit, without needing to stick to the party line.

And there I was on the Spielbergs' property with no screenplay, no short film, no nothing. Nothing to stick in his hands, nothing to lay on his pillow—nothing to make a Hollywood Hail Mary fueled by one part talent and nine parts desperation, the kind of move so many undiscovered talents seemed to need to get noticed.

Yet I knew that if I had pulled a Hail Mary, I might not be fired on the spot, but the ostracization would be so swift and all-consuming that I would die. Maybe not in real life, but in my White House life. In one action, I would be "them," never to be "us" again.

I guessed Mr. Spielberg was lucky: I wouldn't violate his privacy with my storytelling. At least not anytime soon.

Later, in the staff guesthouse, I grilled my agent counterpart.

"What was that golf cart ride about?" I asked.

"Nothing," he said, "Mr. Spielberg was just taking us around. Asking us about our toys. And oh, the Spielbergs are throwing us a lobster barbecue for wheels up."

A lobster barbecue? This was the first time I'd heard of it. Probably because it was just for them: the boys, the happy, toy-loving, gun-shooting boys. Then, out of nowhere, Mr. Spielberg walked in through the sliding door holding a plate of cookies. He approached me.

"Would you like one?" he asked. "They're chocolate chip."

"Sure," I said. "Thank you."

Oh! He was talking to me! Not just talking to me but offering me a cookie! This had to be a good sign.

He put the plate down on the counter, and then he was gone.

Back to playing PlayStation with the agents or creating trust funds for their children—however he typically spent his downtime when preparing to host the president and his family.

The score remained the same. He was a nice guy. And I was just a staffer.

"TEN MINUTES OUT," THE agent said, walking past me at a rapid clip, making last-minute checks. The agent was moving, but the rest of us were standing still. At the far end of the compound, the motorized gate opened and would remain open until the motorcade entered. We would remain still until the president's limo reached the mark—the bit of gravel before the feet of Steven Spielberg and his attractive wife, Kate Capshaw, who stood ready to greet their guests. The hostess looked especially radiant, the picture of casual chic on Georgica Pond, as if it were the most natural thing in the world to wait at attention, outside your home, in your swimsuit.

Eight minutes out.

Soon we'd hear the siren of the pilot car—that false alarm that so many thought meant the motorcade but was not the motorcade, not yet. Standing there, waiting, I marveled at the cloudless blue all around and the sunshine that made you want to forget all of this formality and just go swimming, which would be easy enough, given that the beach was just behind the house.

My eyes traveled back to Ms. Capshaw. There was still time for her to change or to throw something on. She'd have to run fast, though. Few women could pull it off, especially if they were Hollywood. Or, more to the point: especially if they were about to meet the president of the United States.

Seven minutes out. Kate Capshaw was not moving.

I had my sunglasses on, making it easy to stare. I tried to be discreet about it because I was staring at her breasts. The ones you could see easily because Ms. Capshaw wore a one-piece. She had, however, tied a sarong around her hips, with the hem cut at a diagonal; but that wasn't really modest either because when she walked, her tanned legs sliced through the slit. Mr. Spielberg's clothes, on the other hand, were nondescript. A short-sleeved shirt and shorts. But my eyes kept returning to his wife, who was wearing a swimsuit in front of all of the agents, in front of all of the staff, waiting to meet not just the president but his wife.

I heard the siren.

They were coming. They were almost here.

They were here.

The motorcade entered the compound road. We watched the long lengths of metal go slowly around the loop. The spare limo passed. The president's limo approached, then stopped at the mark, right in front of the Spielbergs.

The deputy exited first. He opened the car door for the president. Another agent opened the First Lady's door. They walked over to the Spielbergs, who were walking to meet them. Everyone was smiles and warmth. Ms. Capshaw's shoulders were browning in the sun. One embrace. Then another.

I could only imagine what was running through their heads. Someone had to care that Ms. Capshaw was wearing a bathing suit. Someone other than me.

Then again, maybe no one did. To Mrs. Clinton, Ms. Capshaw might have been just another woman showing off her best bits to the president. Mrs. Clinton had seen this scene before.

So had I. I'd seen my share of rope lines, receiving lines, and meet and greets. I watched the public. I watched the invited guests. How they pulled at him. Called his name. The huggers. The screamers. The wide-eyed gapers. The ones like Monica, waiting.

But everyone wanted him. Men wanted to feel indispensable to his success. They wanted his power to translate into their own. He could deal with the men more easily. He could bat away the ones he had no use for. But the women. I think the women were harder.

Once upon a time, the president was a boy. A chubby boy. A chubby boy of a loving mother, but from a broken home. Now . . .

Being President Clinton meant that every day you encountered women who had traveled from afar to be near you. You were the destination. You were the goal.

Being President Clinton meant you always saw women at their made-up "best"—women using every trick, every available style and cosmetic aid to look as good as they possibly could. They wore new clothes. They wore their best hair. They did this expressly because they would be meeting you. The minute you looked their way, their faces would flower into their widest smiles.

To be President Clinton was to be tested every single day of your life.

ONCE THE CLINTONS and Spielbergs entered the house, Mitchell passed me, said everything was good. If Mitchell was happy, I was happy. I could finally relax.

On a trip like this, with so many body staff, I had little contact with the president or First Lady. I saw their aides, who passed with *hellos*, *looks goods*, and *how you doings*. No complaints. That's what I cared about. Nothing felt as good as that sense of accomplishment, that sense that I'd pleased all the powers that be.

I walked back to the staff guesthouse. There was a large sitting area and kitchen, fully stocked by the Spielberg staff with San Pellegrino, Diet Coke, potato chips, and nuts. Cereal and fruit for the morning. This was where I'd wait until the next departure, and I could get a bit of downtime.

I walked into the exercise room and saw Larry Cockell finishing on the treadmill. I hesitated. I'd seen him more on TV than in person, especially now as he entered and departed the courthouse back in Washington, compelled to testify before a grand jury. But I also hesitated because he was really handsome. I smiled hello. A small smile, not trying to make him talk if he was still working out.

"Are you Stacy?" he asked.

"Yes," I said.

"Donnie?" he said.

"Yes?" Donnie Flynn answered as he walked in behind me.

"This is the RON," Larry said. They were both looking at me. Staring. They looked serious, as if they might be angry.

"So it was you," Larry said, wiping his face down with a towel.

"What?" I said.

Larry stepped off the treadmill. He was not out of breath. He and Donnie approached me. I wanted to laugh and be friendly, but I wasn't sure what was going on.

"So it was you," Larry said again.

"What?" I asked, starting to worry.

"It was *you* who put us in those bunk beds."

"Oh, no!" I said, ashamed as I felt my full culpability in the bunk-bed debacle. "I'm sorry!"

"Sorry?" Larry asked.

"Sorry?" Donnie chimed. "Not just junior twins."

"Bunk beds."

"But I got it approved—"

Larry shook his head, as if I had added insult to injury. "Come now," he said. He motioned for me to follow. I did. I followed the two highest-ranking men of the Presidential Protective Division, who were not just head bodyguards but overseers of all presidential security operations: a mighty framework of investigations, preparations, and executed plans.

Larry opened the door to their assigned bedroom.

Blue and red everywhere. Picture books. Miniature play tables made of plastic. Toy cars, a racetrack. And the two sets of twin bunk beds, one against each wall, that at the very least allowed each supervisor to claim his own set.

"Do you even fit in the bed?" I asked.

Larry gave me a look, which was a look down since he was at least four inches taller than me, and I was five feet ten.

So . . . this was where Larry Cockell and Donnie Flynn were assigned to sleep. By me. A twenty-four-year-old staffer. Out on her sixth trip ever. Watching Larry pick up a toy doll reinforced one of the biggest lessons I had learned from observing my superiors at the White House: never expect power to protect you from indignity.

LARRY AND I CROSSED paths in the compound several times. I was still mortified about the bunk beds, but I liked seeing him. I smiled, said hello, not wanting to seem overeager.

I saw him near the barn-turned-command-post. He turned away from a group of agents. They all stared at me, smiling, knowing.

"It's you," he said, shaking his head. It was a joke now, and I felt happy inside. The big boss was taking time from his boys to joke with me.

One time he even whispered the words "bunk beds."

I smiled and apologized for the twentieth time. I couldn't say "Sorry" enough. Or smile enough. I loved his attention. I didn't know what that meant, so I tried to ignore it. Larry was not the type of man I was typically attracted to, but I felt warm and happy every time I saw him.

IN THE LATE-AFTERNOON SUNSHINE a few agents played basketball on the court behind the staff guesthouse. I watched them. We'd played H-O-R-S-E together the day before. I have a Polaroid shot of me playing H-O-R-S-E on the Spielbergs' court that I still treasure. I'm about to miss the shot, but you can't tell from the Polaroid. I was thinking about joining the basketball game when Larry walked in.

"You know I've forgiven you," Larry said.

"Thank you," I answered, meaning it.

We moved to the kitchen. Bright, modern, gleaming with granite. He stayed with me. He didn't peel away. He didn't say he was busy, that he had to run. Every movement was that of the gentleman. Upright, fluid, present. What a beautiful man he was. What a lucky wife he had to have. And kids. If he had kids, they must have loved their father like there was no tomorrow.

At first we talked about nothing, but then it occurred to me: I could ask him for advice. I had yet to take a self-defense or martial arts class of any kind, but maybe my old economics teacher Mr. Bradin had been right and I should. So, standing there, I told Larry that I wished I knew how to protect myself better.

"Elbows," he said. "Remember your elbows. Hardest part of your body." He showed me how to swing my elbows.

"Let's practice from a choke hold," he said.

Larry Cockell was showing me three different ways to escape a choke hold. I felt nervous, trying to learn the moves, self-conscious to be so close to him. The seams of my dress bodice strained. His shirttail remained perfectly tucked in his khakis. There was no gray under his arms, no sign of perspiration or flush. We changed roles. I felt myself leave my body and watch the scene: there I was, simulating the strangulation of the special agent in charge.

A door opened somewhere. We pulled away from each other, gently.

The MIL AIDE. The one with a queen bed. He greeted us and kept walking.

"I don't know what he was thinking," Larry said in a low voice. "He knew better than that."

"No, I should have said something—"

"No. He should have known better."

Yes, the MIL AIDE who had asked for a master bedroom because his wife was coming up had since put her in a motel. No one told me the full story, other than that staffers much higher than I had protested this breach of protocol. This wasn't a hotel. This was a private residence. Bringing your wife along, uninvited, wasn't cool. Especially when others were sleeping in bunk beds.

In hindsight this all seemed so intuitive, but protocol was intuitive only to a point. If the boss wanted x, you provided x. Power determined rationale. Power didn't even need rationale. The rules of play had to be learned and relearned as players rose and fell on the pecking order.

I headed to the fridge for a drink. Out the window I could see the green ribbon of trees that marked the end of the property. The thicket touched in close to the POTUS guesthouse.

"Do you have to post the forest behind their house?" I asked.

"The swamp, too," he said, joining me, standing on the other side of the kitchen island.

"Oh, that's rough," I said. I felt sorry for the agent post standers stuck out there for hours at a stretch. "Can I get you something?"

He paused, as if considering whether or not to let me serve him.

"A ginger ale. Diet, please."

I found a can, grabbed a glass from the cupboard, and placed them on the island.

"Thank you," he said.

I nodded quickly, then asked, "Do they have thermoimaging glasses?"

"How do you know about those?"

"The TSD agent told me."

"There, press your hand here," the Technical Security Division agent had said to me earlier outside the command post. He'd directed me to touch the steering wheel of the golf cart. I'd gripped the wheel, then removed my hand. "Now look." He'd given me the glasses. In the nocturnal world of the lenses, I could see a bright white patch on the black wheel. He explained how people left energy wherever they went. If someone hid in the forest, an agent using the glasses could see the heat he emanated. If he ran away, the agent could see the heat he left behind. The area would glow for at least an hour.

"I'll have to speak with him," Larry said.

"No, no, no! I'm not trying to get him in trouble! It was just me being nosy."

Larry looked at me, stern as ever.

"Please don't be angry with him," I begged.

He kept his gaze steady for a long moment.

"I guess I'll let this breach of operational security slide."

Agents entered the house, headed toward the kitchen. Laughing, healthy, and back-slappy like overgrown boys. The two men looked at us. I knew they thought they'd walked into something. I chatted them up quickly, trying to make clear that this wasn't a private conversation. I knew agents gossiped. Especially about sex. I didn't want them to get the wrong idea. As Jonathan's mother had said: It's so easy to get a reputation.

Soon enough, Larry and the agents were called away. Pretending not to, I watched Larry as he walked from the room. His beautiful, nut brown skin. His handsome face and his sleek, strong athleticism. But despite these features, he couldn't hide the weariness in his eyes, or in that first moment of walking. No matter his stamina, the weight of this work, in this year in history, had to have taken its toll.

I TENDED TO LIKE white boys. Okay, there was no "tend" involved. At twenty-four, I had always liked white boys. My first crush was on a white teacher, back when I was eight years old and in Grand Blanc, those nine months we lived outside Flint, Michigan. Mr. Dunk. He wasn't even my teacher. One afternoon, instead of walking straight home, I walked to the fourth-grade hall to give him some stickers. And a note, too. Yes, I confessed my undying devotion. I think I freaked him out, but he hid it with an aloof smile.

From then on, one white crush after another, leading up to the man I loved, Jonathan, my boyfriend of both English and Irish descent. I did not pick Jonathan to love because he was white or because he was British. But those were the facts.

Maybe because my father was black and no one was going to hurt me as he did, leaving me for good at such a young age. Maybe because I'd grown up in the suburbs, with all those blondies with eyes that looked like sky over Siberia. Or maybe because I'd taped too many posters in my room of Brit boy musicians, so skinny and pasty and gay; their faces and frames filling my consciousness from floor to ceiling.

All I know is that up until that moment, I had desired white. Black boys scared me. They got into my space. They called after me. They came on too strong.

NIGHTTIME, AND MY ADVANCE TEAM colleagues were out with the road show at the fund-raiser. That was where they should be. I sat on the couch in the staff guesthouse. That was where I should be. Watching TV. Chilling. I'd gotten in a quick nap earlier, and a quicker conversation with my boyfriend long-distance. He wanted to know how the trip was going. Fine, I told him. I stuck to the Spielbergs, the compound. There was no way to mention Larry, even though I was happy that he'd paid so much attention to me, this man, in the middle of his own crisis, taking the time to teach me skills in such a tender way. But I knew that mentioning another man could make Jonathan feel eclipsed. At least it would feel that way to me if the tables were turned.

The motorcade was due back soon. At this point in the trip, arrivals and departures were routine and controlled, given that this was a private compound. Though the RON agent had to stay outside, I needed only to be nearby.

I sank further into the living room couch and looked around the ranch house at its hues of beige and browns. A perfect place to rest, I thought. In the end, all these houses—including the Spielbergs'— seemed no better than hotel suites. Nice, new hotel suites, but hotel suites all the same: designed by someone else, kept up by someone else, but with a cleanliness achievable only by underuse.

But I was glad to be there. Off-duty staff walked through now and then. A doc. A MIL AIDE. Agents. We chatted pleasantly, but about nothing serious as we tried to enjoy the quiet time before the circus returned.

My radio crackled. My serve kit wires for the radio were in my shoulder bag; if I didn't have to wear my earpiece, I didn't. When wired up I felt constricted. I needed a belt for the radio, and after eight hours, my back ached from the weight. With house duty all day and no events to work, I could get away with just hand-carrying the radio—and wearing a dress. Otherwise, I needed pants for the belt and a suit jacket to hide it all.

I put the radio to my ear. *All stations, all stations. Eagle depart, Eagle depart.*

There's a fine line between doing my job and being in the way. I watched from the window. I waited. The motorcade snaked around the loop, lights on in the dusk. I watched POTUS and FLOTUS exit the car, the Spielbergs, too, and everyone looked happy. They walked into the Spielbergs' house together. Clearly their night wasn't over.

Mitchell walked in through the sliding doors, sending in a moist gust of night air. He looked happy. Other staffers walked in with him, a POTUS donor, too. They were laughing, a little loud, a little drunk, maybe. They'd stayed at the fund-raiser a long time, which I imagined to be a swell time, given the Hamptons guest list.

It wasn't unheard of for staff to drink along with everyone else. The night wasn't over for them, either.

Other staffers headed to bed or back to their hotel rooms. The arriving crowd dispersed quickly. Soon it was just me again on the couch, watching CNN.

Then Larry Cockell walked in with the air of success and fatigue of another long day, almost finished. He couldn't call it a night until the principals called it a night. Word was that the Clintons and Spielbergs were having dessert and that it would be an hour or two before they were down.

After a nod to me, Larry walked to his bedroom and closed the door. I hoped he would come back out and talk.

He did.

He was in a button-down and slacks, on his phone. Chinese carryout. "Do you want anything?" he asked me. I shook my head no. He sat down on the opposite couch. I passed him the remote control. At first he paused, then accepted.

I wanted to ask him so many questions. *What is it like to be in charge? What is it like to be in this awful glare?* I suddenly felt greedy for facts. For feelings. But I was afraid he'd think I was milking him. God forbid I should remind him of the press. Or worse: the independent counsel.

"Would you tell me about working with the Chinese?" I asked. "I heard it was a nightmare."

"It was a nightmare," he said.

"How did you handle it?"

He reached for his water. He took a moment to drink slowly, to put his glass down. Even at this late hour, with his serve kit wires exposed and his upper button undone, his shirt still tapered crisply into his waistband.

"Some things are just not negotiable," he said. "You take it off the table. When the Chinese insisted on having their lead security agent sit in the front passenger seat of the president's limo, we simply said no, that's not how we do things, and we took it off the table."

"Oh, isn't that what happened to the VP in Israel?" I asked. "They couldn't agree, so there was both an Israeli and an American agent smooshed there next to the driver?"

"Yeah," he said, smiling a knowing, *not if I had been there* smile. "It's when they sense a quiver, when they sense you're unsure, that's when the powerful will roll you. And we don't like to be rolled. If they don't accept your 'no,' tell them, 'Fine, do what you have to do. But when our president meets with your president and tells him why he was six hours late for the arrival ceremony, are you confident you're going to have a job tomorrow? We can hold our man on the plane. We're lucky, because the president respects our professional opinions—so does the missus. Especially the missus. They listen. Because of this respect we negotiate from strength.'"

Wheel of Fortune was on, but we weren't watching. Outside, the sky was black. The window shades were up. We were on full display for anyone outside, but they couldn't say I flirted with him or he flirted with me or that the evening's proximity would lead to something wrong. We were just two staffers stuck in the same place together, talking.

I plead my own case in my head, yet I knew how harsh judgment could be, and how nuance came off like excuse. You always had to be thinking about your signals. People tried to read one another like stories, and God, they could get you wrong.

LARRY CLICKED AROUND until he hit *Showtime Classic Fights*. Two stocky men stalked each other in the ring and punched each other's faces. He looked to me to see if I would complain. No, this was fine.

A message in his earpiece. Ah, his food. He left and returned with a big brown paper bag, wet on the bottom. I watched him get his food together in the kitchen. He offered to fix me a plate. The noodles smelled good and I accepted, hoping he would stay out with me and not retreat to the bedroom.

He did.

Larry's manners were perfect. Me, on the other hand, I constantly wiped my mouth, not wanting to be messy for one minute. When we finished, he took the plates. I protested. He dismissed my protest with barely a look. I heard him rinse the plates off and put them in the dishwasher.

When he returned, I asked him a question that had been on my mind.

"Are you a father?"

"Yes," he said, smiling.

"Do you have a picture?"

"Yes, I do."

I felt his pride as he handed me the picture of his son from his wallet.

"What a handsome young man he is," I said.

"Are you married?"

"Yes."

"Why don't you wear a ring?"

He paused for a moment.

"So the bad guys won't know."

This was not the answer I had expected, but the minute he said it, I believed his words, for that sounded like the Larry Cockell I knew: he would protect his loved ones, period.

Maybe if another agent had given that reasoning, I would not have believed him, for agents could seem so randy on the road. I felt it, the way they'd stare or joke or call after me in hallways if we were alone. But I also knew that this wasn't the case for every agent. That this man next to me finishing his Chinese food had never been aggressive toward me or given me reason to think that he could have anything but honorable intentions. If he had lusty thoughts, he kept them in check. Nothing about him seemed bitter or curdled or pissed off that he couldn't sleep with every woman around. He paid attention to me, yet I never felt as if I sat before a caged lion. He may have been trained in a thousand dark arts. But I had no doubt that he would always be a gentleman with me.

He could have had no idea how grateful this made me.

Larry bored of the fight. He tried to hand me the remote, but I didn't accept. He flipped through the channels, stopping on a local network. The eleven o'clock news. We stopped talking. We both wanted to see the day's pictures.

I ran up to the TV. Like a little kid, I pointed at the screen.

"There you are, there you are!"

"Don't be silly," Larry said. His walnut eyes were stern, but his stare lost its conviction. He seemed uncomfortable looking at the screen, at the footage of him going into the courthouse and out of the courthouse. Just days ago, the president had spoken to the Girls Nation Class of 1998 in Room 450—the same room my class of 1991 had been taken to for our White House program. The president had spoken proudly of public service and exhorted the girls to do all they could to make our Union more perfect. But after the program, the attendant journalists had peppered the president with questions about the Secret Service testimony. A *Washington Post* reporter who had covered the scene acknowledged in his article the awkwardness of questioning the president about Lewinsky testimony in front of his guests, the young female leaders of tomorrow—any of whom could be in the next class of White House interns.

On the local news that night, there was no great revelation. Just the stoic soldier dragged into Starr's mess. *What did you see? What did you hear? What did you know about the president's conduct? Remember, you're under oath.*

The segment concluded. Larry took a deep breath, shook his head at the absurdity of it all.

He pressed his ear to hear the radio traffic.

"They're down for the night," he said.

That was my cue, too. I needed a few hours' sleep before I had to do my last batch of overnight news clips for the trip. Time to head back to the inn.

We said good-bye.

We didn't hug. We didn't do anything. I slung my bag over my shoulder, headed out to the rental car in the driveway, and drove myself back to Amagansett.

But if I could have written an ending to that evening, the secret ending that only Larry and I would know, I would have stayed there, in the guesthouse. When we tired of TV, we would have gone to sleep. I would have climbed into his upper bunk, and he would have slipped into the lower bunk. And I would have fallen asleep, soundly, knowing that he was below, protecting me.

16

Glass, Girl

Moscow
August 1998

I BROKE STRIDE WITH MY AGENT COUNTERPART, CROSSING OVER the expanse of open dining room space to pull back the white sheers. This was the new Marriott Grand, and I hoped the president's suite would have good views.

The grid, verticalized. That's what I saw. Layer after layer of human activity stacked in high-rises so near that I could see phones on desks and computers turned on and women talking to men. I could feel the energy of strangers in the ritual motion of their lives. This was not the best view, but I imagined the president being pleased to look out and see *life* when he arrived after a long night of traveling, ready for his morning downtime.

The agent walked to my side. "This is awful," he said.

I looked at him and said nothing, for I knew what he saw: each window hid a sniper on his belly, biding his time.

"What do you want to do?" I asked.

The agent paused, looked back out the window. I waited. I knew his recommendations would be grim.

Advance people like us arrived early to assess the spaces where the president would move, meet, and sleep on a trip. The agent worked to protect his physical body, and I, with my list of presidential preferences, worked to protect his peace of mind and to push back against the more overzealous agents who'd prefer to simply seal him up in a popemobile.

But I knew that what fed us could hurt us. Windows allowed sunshine and snapshots of freedom, but they were also holes in the wall, entry and exit points to be secured. They're weapons, too. Agents must think about explosions, of how a nearby blast could turn these windows into sheets of shrapnel blown inward.

We walked into the bedroom. "This is where they'll sleep?" the agent asked. "Yes," I said, showing him my suite diagram and how I had the other bedroom blocked off for Mrs. Clinton's clothes and dressing area.

"I'm sorry," the agent said, "but I need to bring in armor."

No agent had ordered window armor for one of my sites before, but I did not fight him. This was a bad-case scenario for him, for us, and in the end, the president's security was paramount. But as we left the suite, I worried that the president would be angry. Of course he would understand the reasoning, but that reasoning would be poor comfort when he walked into his suite and saw what we did to his living area—how we placed big black freestanding boards in front of each window, with the curtains closed shut behind them. Like precautions for a vampire, I thought, making the suite as safe and dark as a coffin.

I STOOD ON THE INSIDE LEDGE of my hotel room window and pushed against the thick pane that would not open. The coolness of the glass felt sweet against my cheek and forearms, and I angled my head, trying to get the fullest sight of the glowing white Wedding Cake tower, one of seven built by Stalin. Critics derided the architecture, but to me, the white skyscraper looked dreamy, a comic book vision of the metropolis writ real. I knew nothing of the people who lived inside or how infested the walls might be.

I thought of White House life back home. Paul and I were finally in the West Wing, ground floor. Paul had his own office, and I sat in the anteroom with other support staff, including senior adviser Todd Stern's assistant, Jonathan Adashek. I needed Adashek's funny and tart retorts those wretched days when we had sat steeped in other peoples' fear and anxiety, including our own, waiting for whatever might happen to us next. Together we watched CNN's coverage of the first family's Martha's Vineyard vacation in August, right after the president admitted to the world, finally, that he had lied about his affair with Monica Lewinsky. The coverage had felt like a suicide watch. Would the sadness of the Clintons' ordeal finally crush them before our eyes?

Grief still singed my lungs in Moscow, days later, for I had been in the residence the day the president came clean—not just to us but to his wife and child. Paul had asked me to deliver his marked-up draft of the president's remarks to the nation. I walked through grief like radiation from the bomb that had exploded upstairs in their living quarters. I kept my head down. I gave the folder to the aide tasked to meet me and quickly returned to the West Wing.

I couldn't eat lunch. As Adashek and I watched the first family working the airport rope line—smiling for the cameras, trying to look normal, their pain was palpable and it became our pain, dread like lead weight that we had carried around since January.

Don't write anything down. Don't say anything you don't want to repeat in front of a grand jury. Keep it inside you. Keep that frustration locked down hard. Like pain. Keep it inside you so you don't get in trouble. So we don't get in trouble. Every word is a whip to beat you with. We'll

try each one. Some won't hurt. The effect is cumulative. Just wait until we
find the one that really lashes. Then we will cut our names into the fat of
your back.

Some of our colleagues racked up legal bills in the tens of
thousands of dollars. Open the paper, and there were Betty Currie's legal bills, estimated at over $100,000. Maggie Williams, the
former chief of staff to the First Lady who had served through the
Whitewater and Vince Foster investigations, reportedly saw her
bills top $300,000. The Clintons never publicly disclosed whether
they assisted their staff with their bills. At the time, I watched the
president attend fund-raisers for his own legal defense and wondered: Would he look after his staff, too? What would happen to
me if I got drawn in again somehow? The next year could be spent
in perpetual persecution, I worried, with one congressional committee hearing after another led by the chairman Dan Burtons of
the world who had axes to grind. The president's rivals could keep
at this until he was impeached, dead, or drummed out of office. My
$30,000-a-year salary just covered my expenses and did not allow
much for lawyers.

I never wanted to be just another rat fleeing the sinking
ship—as others were called who departed that spring and summer.
I wanted to be loyal.

In 1993, Team Clinton came to town, and we were going to
provide everyone with health care. We could do it. Just *do it.* We
won. What could be harder than winning the presidency? As a
young volunteer, I occasionally answered phones in the Health
Care War Room, in the Old Executive Office Building. The staffers set up elbow-to-elbow work spaces as they had done in Little
Rock. Just like the War Room for the campaign, and we'd won
then, right?

I did not see the secret meetings. I did not hear the debates, the
fights, the mistakes made on the inside. I just felt the push of Team
Clinton's opponents, including "Harry and Louise," the everycouple

from the Health Insurance Association of America television ads who injected Americans with the fear that the Clinton fixes would make things worse.

I saw the dream slip away.

Then there we were in 1997. I sat next to Paul Begala and observed as he pushed his pet policy, the "point oh eight" or .08, the push to standardize, from state to state, what the criminal blood alcohol concentration (BAC) would be. Standardize. The bill provided no additional cops on the roads. No additional alcohol treatment centers. No big actions to minimize the phenomenon of drunk driving. Just brought into line those states that allowed for more booze in the blood, hoping the new strictness itself would lead to fewer drunks behind the wheel.

We lost. The beverage lobby was enormously strong, Paul explained to me, yet so weak as to believe that this bill threatened its business, that the proposal went away and died by the side of the road.

In 1993, we were going to give everyone health care. Now we couldn't pass this. Not a fair summary of the Clinton legislative scorecard, true. And, in 2000, Congress finally passed, and the president signed into law, the .08 bill. But in 1998, with our horizon dark with gloom, this is how I saw progressive prospects in Washington. I sat in my chair like a tar pit, trying to figure out how anyone of good public intent in this town could make more than a dent of difference. I watched Ken Starr speak to reporters about Monica Lewinsky and wondered, what exactly are we doing here? And do I need to be part of it?

The cool window glass felt good against my face, but I was full of riled-up energy. With the dance videos on the TV on full blast, I jumped from the ledge to the bed, ready to be out in the world—and out there now.

Detroit, Michigan
1978

THE MOTHER CREAKED OPEN the door and sucked her breath. Startled, the girl lost her grip on the windowsill and stumbled backward.

"Why aren't you in bed?" the mother demanded. "Why are you by the window? You can't open that!" The girl felt each step in her belly as the mother lunged for the glass and pushed it down.

"See?" She reopened it enough to stick her hand outside. "There's nothing here to protect you. If you slip, you fall all the way to the ground."

The girl felt the flick in her throat as she turned to look at her. Her mother wore white, ready for work. Quickly she closed the window and turned the lock.

"Do you know how lost I'd be without you?"

The mother grasped her head. The girl loved the feel of her fingers, her palm smoothing down her dark, wispy hair, how she bent down and kissed each eyelid. One, two. The girl smiled open-mouthed, wet-lipped.

The mother put the girl back into the small bed, near the larger one.

"No," the girl said. "Not sleepy."

"You will be," she replied, pulling the sheet up to her chin.

"Too hot," the girl said, kicking it away.

"Hold on." The mother left the room. The girl, listening, heard her father say something. Her mother said something in a strong whisper, then returned. She had the fan in her hands. She knelt, plugged it in, angled it so the wind hit the girl's body.

"Stay in bed now, okay? I'll be back soon."

The door closed, making the room dark, or as dark as the city allowed. The girl felt the ache settle in her belly. She always felt it when the door shut and her mother left for the night. She wanted to run after her, to go on the bus with her, to sit at the nurses' desk and color her pictures.

Why couldn't she go to the hospital, too?

She kicked at the sheet till it bunched at the foot of the bed. She grunted and turned, her exhales little huffs, until the wretched energy seeped from her body and her eyes started to close.

The pillow was hot. She turned the pillow over and pressed her cheek against the cooler side. The coolness slipped away, and she tossed again and she turned. The sheet touched her foot, and she kicked it to the floor.

She sat up and faced the fan. Too far away, she thought, as she slipped out of bed and reached for the big grilled circle. The whirring was loud. She opened her mouth. *Ahh-h-h-h.* That made her giggle. *Ah-h-h-h-h-h-h!*

Smelling the metal, she leaned forward and pressed her tongue against the grill. She ran it slowly across the ridges. The dirt was gritty, but she didn't mind. The air cooled her tongue, tickling the back of her throat. She stuck her tongue between the slats.

"Ahh!!!" she screamed. She yanked back her face. Her mouth hung open. She was still. Startled. The pain flashed hard in her mouth.

The girl wailed until she choked on her breath, a scary spasm that stopped her from crying. Her breathing settled and the panic softened, but her eyes and cheeks were wet and getting wetter against her hot skin.

"Daddy!" she howled. No response.

She stumbled up and padded to the door. She opened it and followed the light through the hallway and into the living room.

"Daddy?"

He lay slung in the La-Z-Boy. He made sounds out of his nose and mouth. She ran and grabbed his limp hands, hanging from the arms of the chair. They were warm. She pressed against him, her small sternum against his knee. At their feet were two bottles. One, she knew, was sweet, like red pop. The other one burned her mouth.

The father's eyelashes opened slowly. She loved his warm brown face and his big brown eyes. She pulled at his shirt and climbed up on his lap.

"Bella," he breathed, his eyes sinking shut again. He wrapped his arms around her slowly. "Why are you up?"

"I hurt my tongue," she said.

He opened his eyes. She stuck out her tongue.

"Does it hurt?" he asked.

"Yes," she answered, even though the pain had worn off already.

"I'm sorry, baby."

She closed her mouth. He held her to his chest and closed his eyes again. In less than a minute, she heard his heavy breaths. The girl loved to be held by her daddy, but she was too hot. There was no fan out there, and his breath was sticky on her neck. She loosened herself from his grip and slipped to the floor.

Legos. A Raggedy Ann. There was not much to play with in the living room. Her mom had brought her an invisible-ink pad from the hospital, but she could not find it.

The windows ahead were two tall eyes. At night, their shiny blackness was like nothing else in the dark apartment. Jammed up, the left window stood two feet open, as far up as it would go. She could stand on the windowsill and press her face against the glass, and her body would feel the outside cold.

She walked over until her thighs bumped against the windowsill. The cracked white paint scratched her fingertips as she stuck her head out into the darkness. So cool. So refreshing. She turned her face to see all of the buildings, the near skyscrapers of downtown. A patchwork of light and dark. Below, the glow of signs red and pink, the streetlight changing from green to yellow. Two men called out to each other, the sounds of their words rising up like the steam off a boil, then the boil itself.

Someone was cooking in another apartment. She tasted the sweet sauce that comes with ribs, the sauce she liked on bread. Music floated up, too. She loved the music! Music like she heard on the radio or her father's records. This time, the woman singer sounded happy.

The girl climbed up on the sill. She reached for the raised window and straightened her knees until she stood tall, her face pressed against the glass. Her belly and toes felt what they had ached to feel all along: the relief of the air outside.

The girl gripped the wooden bottom of the window frame. Her sleeping shirt had ridden up, and air tingled her belly. She stuck her face against the glass, squishing her nose, and looked down. Now she could really see to the street below through the glass.

She jumped backward, stumbling on her behind but quickly uprighting. She went for her Legos and grabbed a red one. She took it back to the window and, with all the force she could muster, threw it into the black air.

"Come back here, baby," called her daddy.

She ran across the room and bounded into his lap. He opened his eyes slowly, but wide. He had a very sweet smell, the smell that was his. She hugged his chest, burrowing each hand around his sides.

Nighttime in the apartment. Only the grayish midnight light from the outside, her forehead becoming moist again from the heat.

"I want to ask you something," he said. "Who would you rather live with: me or your mother?"

She looked up into his face. His deep brown eyes looked into hers. She felt her small heart beat next to his and felt no separation of flesh. Her answer was as involuntary a response as her next blink.

"You, Daddy," she said.

He hugged her tightly, then kissed her on the tip of the nose, then on her lips. She was so happy. She kissed his lips again. The pressure lasted a long time. He squeezed harder. She tensed. She could not breathe well.

He released her. She was relieved but quickly wanted his squeeze again. She tried to kiss him again, but he pulled away.

"I want to be with you, Daddy," said the girl, and she meant it. But her stomach started to hurt. The question had created a painful tug-of-war in her body. She held him tighter, but his grip did not change. He kept her there, his hold looser this time, and she fell asleep.

When she opened her eyes, her sight was blurry, but she saw the gray rectangle of the window shade hanging before her, haloed in white, and she knew she was back in her bed. She looked over to the higher, bigger bed. Her mother's chest rose and fell slowly. The girl closed her eyes again, falling back into the deep sleep that was possible only when her mother slept near her, watching over.

Moscow
August 1998

A FEW DAYS INTO OUR TRIP, Russia plunged headlong into a financial crisis. I tried to withdraw money from ATMs and could not. Neither could my teammates. I was okay, and so were they: we all had enough cash and credit to cover our expenses. The only sad thing was that I was forced to forgo my purchase of a samovar for Aunt Ella, my mother's Ukrainian-Jewish friend who had fled with her husband some years back and had never been able to return. I did not have the dollars.

Power manifested itself in currency: how strange to eat meals in the hotel restaurant and finding the prices printed in U.S. dollars! The restaurant manager explained that the ruble fluctuated too much, that they'd have to keep reprinting the menu if they kept it in rubles. In Mother Russia, the U.S. dollar represented stability. No need to say we won, I thought: victory was printed in the cost of the beef Stroganoff, in the smile of the street vendor when the bills hit his hand. Or the ease with which we gave away our dollars, not believing that the source of our wealth could ever run dry.

Watch us as we walked those Moscow streets. Teammates wore polo shirts with CLINTON-GORE on them, announcing to the observant that we were the kids from America come to arrange the president's trip. Kids. Disciplined when it came to job execution but not security. Watch us with our cheeks plastered to our cell phones, doing a piss-poor job of not talking about event sites and presidential movements, the details of our jobs that were presumably fine dining for anyone listening in, for anyone who desired to disrupt our events or hurt our president.

We walked with talky braggadocio, never having lived through national disorder or war or the consequences that come when security is compromised. Meanwhile, the Secret Service agents repeated two words at every countdown meeting: "Operational security." They gave their short speeches: *Watch your papers. Watch your cell phone conversations. Use the safes in your hotel room. Use the burn bags.* Yet how many times had we heard this? I listened and I did not listen, and out we went into Red Square with our Clinton-Gore gear past St. Basil's Cathedral, with the insouciance of the protected.

At one countdown meeting, an agent was speaking at the dais, and despite my best intentions, I tuned out his speech, just as I did when flight attendants gave instructions on how to find the flotation devices beneath the seats—information I knew might be crucial someday but just could not hold my attention. This time, the agent announced that the embassy's regional security officer

(RSO), Tony Bell, had a few words to share. I kept working on my to-do list for that evening.

"You see all those Mercedes sedans in the streets?" the RSO asked. "If one of those almost runs you over, do not react. Keep on walking."

Huh?

Low laughter erupted all around. My head jerked up, and I met my colleagues' eyes. What kind of advice was that from the Americans' head security officer? Look at him, I thought. The RSO was a strapping black man in his fifties, the kind of man you wouldn't think would be afraid of anyone—or at least wouldn't let you think he was. I had already seen plenty of Mercedes sedans rushing down the streets, and I had taken that as simple confirmation that there was a new order in town, the West had really won.

The RSO explained that the guys in those cars were usually mobsters and that the best strategy was to refuse to engage with them. The U.S. embassy felt the need not only to alert us to this "problem" but to advise us to keep on walking.

I tried hard to keep listening, but my mind was already back in the Moscow streets, lost among all the summertime people and the places we'd seen so far. The last thing I had felt was scared.

The night before, a group of us had gone to the Bolshoi Theater. Staff and agents together, on a lark. Ticketless, having been told that we could get scalped seats at the front steps. Fifteen U.S. dollars each for second-row center seats. *Aida.*

To feel the history as we walked inside that dark, storied hall. See the Russians—this was theirs, their pride. Men and women around us who may have been rich but did not look ritzy. Men and women who may have peopled the national cultural brain trust, dressed so carefully in their blacks and browns. Even the most matronly women returning smiles when I looked at them.

Look down into the orchestra pit. Such gorgeous men in their tuxedos the luscious color of the mobster sedans. Men in their prime, ready to raise their bows. Above them, the drawn red curtains. Gold hammers and sickles stitched into thick velvet. The old state symbols may have remained, but soon the curtains parted.

The Russians wore their brown face paint and set out to replicate the exotic splendor first brought to the Italian stage by Giuseppe Verdi in 1872. We sat rapt as they sang potently of love in ancient Egypt, bringing to life the woman, enslaved, who must choose between her warrior lover and her father and their embattled native land, all the while suffering the jealousy of her mistress, who loves the warrior, too. Forget that the lovers are doomed—or because of that, listen to their beauty. Feel the music pierce. By song's end, we were all connected by the lit specks of God in our hearts.

I wouldn't let the RSO's message frighten me. I wanted more of this feeling—and to give it, too. I felt pierced when the president or First Lady spoke. But I knew that the White House was no longer the place for me to learn or seek joy. I had new frontiers to explore. I wanted to live the artist's life, not the politician's life. I wanted to be free like that. And if a Mercedes sedan almost runs me over, that was okay. I knew to keep moving. Not to let the trespasses of aggressors stop me in my tracks.

THE BLACK BOARDS LOOKED UGLY. I stood in the president's suite and beheld each tall freestanding board the color of pitch that stood before the dining room, living room, and bedroom windows. The sight took my breath away. I felt like this was the mise-en-scène of a psychological thriller where the scientist punished his captive by blocking his light, and therefore his life.

My agent counterpart had warned me, so I came prepared. The embassy had helped me secure art from a local school. The students had taken woodcuts, pressed them into ink, then pressed them against white paper, making images of St. Basil's Cathedral and other local churches and buildings, the beautiful sites the Clintons would not visit on this trip, given their tight schedules. Their pictures looked cool. I had more than twenty of them. In the short time between the security sweep and arrival, I taped them to the boards, creating a strange chessboard effect. Perhaps uglifying the place further. I couldn't tell.

I checked the curtains. The sheers were pinned shut! I felt the pins and sighed hard, hearing the president in my ears: *It's like a goddamned cave in here!* The agent and I had agreed only to the armor. On later trips, the agent would be told to unpin the sheers, sometimes by the president himself. But in Moscow I stepped back, shook my head, and hoped that the president would not freak out when he saw what his suite looked like.

The amenities waited on the dining room table. Caviar, just set out by hotel staff. European bottled water and Diet Coke on ice. Wine. Crudités in the kitchen. Everything arranged ahead of time with my hotel counterparts, for hotels always provided impressive amenities for the president, free of additional cost. I suggested the drinks and vegetable plates, while they insisted on the caviar and the ice sculpture centerpiece: a large tenor saxophone, perfectly cold, without a single drip of water or any moisture at all on the curves of the horn.

I knew the hotel had fussed over the saxophone, trying hard to impress. But all I could wonder was: How many times had the president seen *this*? Saxophones, saxophones everywhere—on cakes, on cards, on others' lapels. When a powerful man let the world know he liked something, it was mirrored back to him in a thousand ways.

The president appreciated kindness, I reminded myself. He was not a snob concerned more with appearances than content. He would eat the gourmet food and be happy, able to nap before going back to work and doing what everyone expected him to do: lead.

But I still worried about the suite. Environment mattered. If I could help him in one way, it could be to help him keep his peace of mind. Help him be happy so he could do his job. That's what we all wanted. To be of service. To be useful.

On arrival the president seemed tired and not inclined to talk. He and Mrs. Clinton walked behind me as we approached the suite. We entered. This all happened fast. I pointed out which bedroom was Mrs. Clinton's dressing area, where the thermostat was hiding, much as a bellhop would; there was never hotel staff present when he entered his suite. I saw him eye the dining-room table and give a nod of acknowledgment.

I waited another moment to see if I would receive a request or an instruction. The president or the First Lady might still react to the windows. They might still complain bitterly. The president was famous for his red-faced rages, and I braced myself. He had never yelled at me, but there was a first time for everything. If I thought the suite looked bizarre and forbidding despite the taped-up art-work of children, so might the Clintons.

Nothing. No opinion uttered one way or another. I overheard the body aide say they were down for the morning, and I took that as my cue to get downstairs to help with the arriving staff—and know that once again, all our hard work resulted in what we wanted: a smooth arrival for POTUS and FLOTUS, one with the comfort and continuity that allowed the president to focus more fully on the problems we elected him to address.

LATER THAT NIGHT, I stood at my hotel room window again. I could quit now, I thought, and go straight to California, that place Americans run to when they want to reboot their lives. Just do it. Stay with Christina and Melissa, my friends from high school, until I got on my feet. Begin a life of writing, a career in entertainment, touching hearts and minds if I got any good at it. Break things off with my boyfriend, too. Do it all at once. All summer Jonathan and I fought. Breaking up, reuniting, breaking up, fighting. Enough was enough.

I looked at the Wedding Cake building, glowing white in the night sky. A monstrosity to some but the tip of a dreaming spire for me.

When I returned to work, I gave Paul my two weeks' notice.

III

AFTER

17

The Trouble with Power

Outside Abuja, Nigeria
2001

WE STAFFERS STOOD PROUDLY NEAR OUR PRESIDENT IN THE late African afternoon, enjoying the last event of the trip. In a small village, dirt paths swept as neatly as a just-cleaned house, we stood shaded by the party tent erected for the occasion as the local governor and president sat in the chairs of honor. I still called him the president, as did everyone else. But this was 2001; the machinery that had once supported him, and us, had shrunk so much it could fit in the palm of his hand.

Village dancers performed. In the circular clearing, women with pretty brown skin danced their synchronized stomps and shakes that went rapid, then slow but remained controlled. The drummers kept time. The entirety of the village surrounded us, sitting as the big men's audience. We were in the governor's home village, and for days before, village life had been upended to prepare for this moment. Everyone seemed awed and honored to have the presi-dent there.

Look at the president on his throne, smiling, too. He might be "former," but he was still formidable: his powers to charm, cajole, blew away anyone; his brilliance remained unrivaled. If he was still the magician, he did not need us for tricks; he needed us to manage the show. He needed staff, including ex-staffers like me who served as occasional volunteer advance, to engineer his public moments so he was not overwhelmed by well-wishers, fans, planners, and schemers. He needed us to shepherd in those he wanted to see and keep at bay those he did not. I was able to help on this trip because I happened to be living in Abuja, in the Julius Berger Nigeria (JBN) staff compound, for I was engaged to Kai Aab, a German master stonemason and engineer who ran the JBN natural stone factory, a man I had met while doing the president's advance in Abuja in August 2000, right after the Okinawa G8. Though the balcony experience had spooked me, I soon realized that no one was going to treat me differently, including the president. Thank goodness, because if I had not said yes to the next trip, I would never have met my husband-to-be.

The women kept dancing. The drummers kept time.

Two hours before, I had witnessed the president at his most powerful—and, to me, his absolute best. He had attended the African HIV conference with African heads of state. Important discussions ensued, yes, but the most powerful moment for me was the picture he created with a teenaged boy, born infected with the virus.

I stood backstage with Christina,* my advance lead, an impres-sive woman of West Indian descent who was a master of details and perhaps born with dollops of drill instructor in her blood. They had sent her to Nigeria as the advance lead, to do the work a dozen folks

---

* This name has been changed.

would have done the year before, and she never blinked. I tried to help her, now that I lived in-country with my fiancé. But I was no Christina. Site advance, press advance, travel office, motorcade—like a good lead, she knew how all the moving parts worked, while I had rarely strayed from RON territory. I watched and learned and advised where I could.

A teen boy, Abayomi Rotimi Mighty, approached the lectern. He spoke of his life with HIV. How he had been born with it and how he'd been spurned by neighbors, by strangers, even by members of his family, how people believed they would catch HIV by touching him. How his life had been very hard. Still, he had not given up. Still, he was happy to be alive.

I started to cry. I knew I was not alone. The audience clapped loudly for him during his speech. The president walked onstage and hugged him, long and hard. *Click click click.* Moment after moment elapsed, and the president wouldn't let him go. *Click click click* as the boy beamed and so did we, as the cameras captured this love to be shared with the world. How that boy must have felt the president's acceptance, his utter lack of fear that HIV could somehow rub off on him. The president held him by the shoulders and said laudatory words about the brave young man who had stood up in front of all these leaders, in front of all these cameras, and told us about his pain and about his hope.

The power of the image. How one moment captured, disseminated, could shatter verities and shift human hearts. If one of the most powerful men in the world was not afraid of this boy, was not afraid of being *contaminated* somehow, how many more may be moved to stop making the heartbreaking decision to shun the HIV-afflicted in their own communities, in their own families?

Every Clinton staffer had moments when all the work, all the barked orders blurred to background and in the foreground stood, in the most clarified light possible, the reason you supported this man. This was such a time. A single action, captured in image, piercing consciousness. If the will to power versus the will to love fought it out every day in the president's conscience, I never doubted that love knew how to triumph. I watched the president smile at the drummers, at the governor, so clearly happy being with all kinds of people, no matter how different they might outwardly seem. The comfort of a man who accepts others for who they are, the comfort of a man full of understanding.

I sat in the Rose Garden, too, one day in 1993 for the signing of the Oslo Accords, when the president stood wide-armed and harboring behind Yassir Arafat and Yitzhak Rabin, as they extended their hands to each other and shook. *Click. Click. Click.* We witnessed those gestures, those single moments, captured, that set off flashbulbs in a million hearts, illuminating the way of peace. Illuminating. Not forcing. Nothing can be forced except a bullet through skin. An assassin did exactly this to Rabin two years later. The need for power, the need for security, pushed one man to cut down another.

I looked at Doug, the president's body aide, standing in his dark business suit out there in the African bush. He was the alter ego, the man who spoke for the president's needs and wishes. Justin stood nearby. He was the young man tapped to replace Doug, who was rumored to be headed to law school. Justin seemed caught in the unenviable position of trying to show he could play Doug's role, while contending with the fact that Doug was still in control. Justin worked hard and endlessly, and he would later become one of the president's most indispensable aides. But he often had brusque words for me, and I had yet to catch him smile.

My attention turned toward the Secret Service agents. The president no longer had the same level of protection he once had. I watched the agents ignore the dancers and scan the perimeter, then the seated audience. With the event in motion, I could admire the power of the dance, the pleasure of submitting to the fact that we were guests in the village, in the middle of an unfamiliar landscape.

The agents, however, must remain focused on the possibility that at any moment, one of the hospitable people around us could suddenly turn violent. They whispered on their radios as the president clapped, enjoying himself.

Walk some feet up the freshly laid path to the main road. If you had shouted out into the distance, who would have heard? A few passing Nigerians, maybe, if there were any. We were so far in the country, so far from the capital. Nowhere near "backup" if backup were somehow called for. In America, we relied on our web of police, of military, of concentric circles of security. In the village, we must depend on our hosts. Hosts who were not us, who were not part of our group.

This after the near riot. A riot I did not witness but heard about as soon as the president was seated. The governor and president had left Abuja together in President Clinton's vehicle, headed for the village event. Apparently the governor had asked the president to step out and visit a busy, crowded outdoor market. For us, this was a nightmare scenario: markets, given how packed and unpredictable they were, could easily descend into chaos. But the governor asked him to walk to the market for a moment, and the president said yes.

People saw the motorcade approach on the sun-parched road and stopped to stare. The president stepped out of the vehicle and onlookers rushed toward them. There were only so many agents. How easily this could have gone wrong. The president shook hands, and with every second that elapsed, more people rushed up, and there was a point when the staff and agents knew he must return to the vehicle. They slipped POTUS inside, a retreat that must have felt like cheating the cresting tsunami wave that wished to swallow them.

Like a *love surge*. That's what the Reverend Jesse Jackson had called the rush of the massive crowd assembled in Ghana to hear the president, back when the TV news cameras had captured the president yelling "Get back!" because they were crushing two women against the barricade fence. This during the president's first official trip through Africa, during his first term. How scary it looked on TV, to see all that pressing and unfettered desire: the strength of the entire body, multiplied by thousands and thousands, focused on reaching out and touching him. Clinton was the powerful one, the ultimate juju man who knew the secret for ruling billions of other men built just like him.

No matter the motivation, a flood of men and women focused on one man could be overwhelming. Especially if you must guard him from harm.

The motorcade stopped again at the local state line. No one on our side wanted the motorcade stopped. The advance lead, Christina, had led a strenuous fight with our Nigerian counterparts on this for days. I had watched her. I listened to her calls. Once they crossed into his home state, the governor had wanted the president to exit his vehicle and enter the governor's car for the rest of the drive to his village. They stated that protocol dictated this, but our side did not care. We never agreed to it. Christina stated that this would not happen, that the president would remain in his vehicle. "Do not stop the motorcade," she told them. I heard her. Christina was formidable herself. But this was a contest of wills at all levels. The president would not leave his car, she warned. *Do not stop the motorcade.*

When the governor's vehicle stopped ahead of them, Doug and Justin were furious. Why were they stopping short of the destination? This was, by definition, creating vulnerability. The staffers jumped out while the president remained inside, talking on the phone. Doug and Christina told the Nigerians the president was not switching cars. Back and forth, back and forth, a standoff behind sunglasses, the asphalt burning beneath their dress shoes. The president took no obvious notice of what was happening outside. Staff and agents stood outside sweating, negotiating. A crowd grew bigger, nearer.

The president stayed where he was.

The governor gave up. All present reentered their vehicles, and the motorcade drove on until it reached us, waiting at the village, and the tale was recounted.

But not before Doug yelled at Christina that this was all so disorganized. Not before he asserted that maybe she was working for the Nigerians, not the president, having privileged their wishes above his. She told him she was not rogue, that the governor had done this on his own and that Doug should know this. To hear her tell me that hurt my heart. I had watched how hard she had worked all week, sent there to do a job that, just the year before, a full team would have handled. I had been asked to help her, but she had done all the heavy lifting. She was the Lead. And to the outside observer, the show not only went on but was a ridiculous hit.

As now. The performance continued, and the president looked pleased. For us, there's a certain peace you feel when an event is under way and there's nothing more you can do as an advance person one way or another until the principal is ready to depart. Nothing else to do but watch politely, knowing that the president would soon be wheels up and we advance people, staff, and agents could finally relax.

Except for Doug. He kept going with the president. He managed the next day of events, then the next day, wherever the president landed. I looked at him in his suit, and I could not help but admire him—despite the hurtful things Christina told me he had said. Staffers at this level were expected to engineer miracles on a daily basis, imposing order on chaos no matter the love surges pushing against them. Doug managed to do this every day of his career. He sat in the African heat in his suit and he did not sweat. I respected Justin, but he was still so new; he would need time and experience to know how to run this man's complicated life.

"You know," I said, leaning in to Doug, "Justin is no you."

Without missing a beat, Doug leaned in to me. "Christina's no you," he said. "You're one of us."

This stopped me on a dime. Was he kidding? How in the world was I somehow one of them, and Christina not? Did he not see how hard she worked?

You're one of us.

I thought about such belonging. Maybe his declaration was only a throwaway line, but I felt myself touched by it. Doug uttered what had always been my deep, unspoken wish. To truly belong, safe and happy. I had always thought of that in terms of another man's heart, but clearly, by extension, this meant the village, too. My only intent had been to stroke Doug by saying something I believed true, for at that point Justin *was* no Doug, as he was just learning the ropes, but that was the kind of thing you said to someone about to be replaced so they wouldn't feel so . . . *replaced*. I had not been gunning for reciprocity. I wanted to stop and deconstruct, but Doug continued on, saying that he had tried to get Christina taken off the trip. I listened, betraying no emotion because I now took his words as a blow. Every day of this advance, I had watched Christina push a ship up a dry hill with her shoulder, in a country where the best old-hand advance people got rolled. Telling our hosts that our president wanted things one way was never enough. It was constant pushing and pushback, day after day, where the brinkmanship literally went so far as waiting at the side of the road for the president to exit the car and the president refusing to do so, in a country where embassy officials shook their heads when plans went awry and used the acronym "WAWA" as an explanation: "West Africa Wins Again."

The women shook faster as the drummers built to their crescendo.

You're one of us. The words troubled me more than Doug could have known, for I sat there asking myself: How had I become part of the group but Christina had not?

At that moment, I knew something, an idea I'd always understood rationally but now was felt knowledge. There was nothing *I* could do to become one of them. They had to choose me. They had the power. They had the power to give, and they had the exact same power to take away. I could run marathons, carry cars on my back for them, and that would never force their hand. The power was with them. And because of this, I could be expelled as well. The scary realization was that I could cease to belong at any moment of their choosing. I could be marched to the village edge and told to fend for myself in a wild land, alone.

I could be abandoned, again, as I had been by my father.

Unless, of course, I mastered the ways of power myself. Some politicians and staffers lived accomplished, even heroic lives this way, and they amassed, aggrandized, asserted control over others when necessary. Some operatives took it farther, doing enough dirty work for the bosses that they wanted, or needed, to keep their positions secure. They became masters of secret keeping, intrigue. But that too was a dangerous game, a hateful game really, to have to keep looking over one's shoulder, always hyperaware of one's environs, to live like Vice President Dick Cheney or a sidewalk mafioso, with your secrets like plaque building up fast inside you. Then one day the powerful deem you too dangerous or the up-and-coming want you gone. And gone you'll be. Gone, just like that.

I sat under that African tent, the swept dirt beneath my feet, thinking that I had left government because I feared that power, unlike love, would use and discard me. Power was scared. Power was insecure. Power was not love. Yet to love we needed enough power to keep aggressors at bay, to neutralize those who believe they must exert control over us to feel secure, to simply survive. My new life in the JBN compound, in Kai's protective arms, was the most tender love I had ever known. But the possibility of violent crime scared me in Nigeria, as it did so many of the other expatriates. So much so that we slept in a compound surrounded by razor wire. I did not want to live this way. Eventually, I would think this life unsustainable, and I would ask Kai if we could return to my homeland, America. But I knew that our early happiness thrived because we lived with bars on our windows, with our own police on patrol. We had the power necessary to feel secure enough to unclench. That's the trouble with power: how can you have love without it?

The women finished their dance, and we clapped. Nearby, the agents scanned the perimeter, waiting for sudden moves.

18

Government Girls in
the Age of Obama

GOVERNMENT GIRLS: IN THE 1940S, THIS IS WHAT THEY CALLED the young women who migrated to our capital city dreaming of service and wishing to better not just the nation but themselves. They worked as secretaries, as analysts, as voices on the phone when citizens needed help. The World War II effort created a great labor vacuum for young women to fill, and they seized the opportunities, often doing so without female mentors or set paths to follow. In many vocations, they were the first of their kind, shaking up the status quo by their mere presence. As Megan Rosenfeld wrote in 1999 in the *Washington Post* about her mother, who was one of the original "government girls": "The women changed Washington in ways they were largely unaware of. Their youth, competence, energy, enthusiasm and problems were sometimes disconcerting to the wider world, unaccustomed to working women but unable to manage the war effort without their help." These trailblazing women often scandalized the sensibilities of "unaccustomed" Washington. Rosenfeld continued: "[Government girls] were chastised for failing to wear hats and gloves at all times, for letting their hair 'go at loose ends,' allowing the seams of their stockings to be twisted, failing to freshen their lipstick enough. No less a personage than a Washington mayor, one Fletcher Bowran, told the D.C. Council he was distressed by the number of women he'd seen wearing slacks."

I read that indictment and had to laugh, for it reminded me of my visit to Vernon Jordan's office and the issue of whether or not to wear hosiery. There is power in appearances, no doubt. But when you have a mission, when you have important work to do, strict adherence to style over substance can seem like misspent energy. Especially when you're fighting to stay alive, as we were in the Clinton administration, as my peers were who endeavored to be so audacious as to support Barack Obama in his risky run for the presidency—this in a nation that for so long had operated under the axiom that blacks were not supposed to have the best of anything, including the World War II jobs that went to inexperienced white women instead of capable black men.

Despite inequalities, no one can deny that there has been true, measurable social progress in our nation. As a young black woman who entered government service and thrived and stayed till I had my fill, I am living proof. Young female staffers of all colors have been able to prove themselves to be reliable, dogged fighters. Put us in charge of policy development, or in charge of an event or a mailing list, and the rest of the world can fall away as we laser-focus on our tasks. When I think of advance work friends like Anie Borja, Angela Baker, and Lisa Starks, all of whom either staffed or volunteered for major political figures in the 2008 election cycle, I see women who take deep pride in achieving tough goals. Whether our makeup was on perfectly or not was never the first consideration.

* * *

ON ELECTION DAY 2008, the dream goal was attained for Obama staffers, including the armies of young women who had phone-banked, canvassed, and run events for both Senator and Mrs. Obama. Several former Clintonites staffed the Obama campaign, including Christina, the advance lead from President Clinton's 2001 Nigeria trip. I changed her name in the previous chapter to protect her identity, so as not to draw unauthorized attention to her as she devoted herself to the Obama effort and now to the Obama administration.

I am proud of Christina. This was the same staffer whom Bill Clinton's top aide declared was not "one of us" while we stood together in that Nigerian village. I had been mortified by his statement, having watched Christina perform magic holding that trip together, her grit more potent than any pixie dust. But in hindsight, maybe the aide was right: Christina joined Team Obama early and became a full-time advance-team lead. She flourished, becoming known for her strict, no-nonsense style. Her teams were not your father's advance teams: Christina cracked down on obnoxious behavior, including foul language and wild partying that, if reported on by observers, could reflect badly on the candidate. Every presidential campaign cycle, journalists seemed to file stories about "bad attitude" advance people, many of them scorched-earth types who treated people rudely in cities they thought they'd never return to, who told cops who pulled them over, "Don't you know who I am? I work for the campaign!" Christina made it clear that no such arrogance would be tolerated—arrogance that a desperate opposition would attribute to the candidate himself.

No idol adoration would be allowed either. Christina rebuked event volunteers who acted more like fans than teammates. Obama experienced mini–"love surges" at so many stops that staff and volunteers had to work hard to keep his areas controlled; no easy job when your candidate is becoming a physical law of the universe unto himself, attracting people as if he were a human magnet. Christina kept an eye out for volunteers who forgot their roles and allowed these onlookers to flow right past them.

"Special forces"—that's what Obama nicknamed Christina. A high-ranking staffer once pulled her aside and said that whenever the candidate learned she was the lead on a trip, for a moment he appeared calmer, assured that his events would have structure and order. A candidate like Obama needed strong-willed staffers who knew how to create open, safe spaces for him to enter where he wouldn't be overwhelmed, where all his energies wouldn't have to go into upholding boundaries. Christina provided this for her candidate. Maybe she got rolled a few times in Nigeria, but, as I said before, that was a country where even the best veterans got rolled. Clearly, Team Obama saw what it had in Christina. When I learned of her new campaign life, I felt a vicarious vindication. Every trip I worked with her, she tried to be the best soldier she could be. She had found her right army in Team Obama.

She was one of them now.

"I LIKE BARACK," said a black woman interviewed on television, "but who I really love is Michelle . . . imagine, the sight of two black children playing on the White House lawn? Imagine, a black man and a black woman so clearly in love, so strong together, as our president and First Lady . . . imagine . . ."

That's what we supporters did. Now the picture of young and female in the White House is two lovely black girls, ages seven and ten. In March 2009, the Obamas purchased a swing set for Sasha and Malia and had it assembled on the South Lawn. With that, another collective wish was fulfilled.

What a message this sends to the world and, more important, to the little black girls and black boys watching, too. Travel from inner-city Detroit to suburban Troy, tour the schools, and ask: Why shouldn't inner-city children feel angry or ashamed that the suburban kids get the best buildings and books while they make do with less? If they live in chronic crisis, either at home or in their neighborhoods, and never see much black life beyond their few blocks and BET, how can they be blamed for internalizing the message that their options are limited, that unless they can rap, sing, or play sports, they must be less than? Now we can watch the Obamas and their girls and see another truth: black children are to be cherished. They are to be given every opportunity. Legs up as the girls swing, as Malia and Sasha pump themselves up higher. This is now the picture of our girls, the daughters of the nation's black first family. A black Detroit girl may not have that life yet. But she can watch them and know it's attainable. That if she can't have it for herself, she can make it for her child. The same month the Obamas built the swing set, Mrs. Obama sat with students at Anacostia High School in Washington, D.C., and declared that there was nothing "magic" about her story. That she wanted the students to know that her parents were "working-class people." No child who watches the Obamas should believe that his or her blackness excludes him or her from the American dream. You just need love, the kind Mrs. Obama's family provided her, the kind the Obamas so clearly provide their children. And knowing that the president was raised by his single mother, we know that those of us without both our birth parents can still be loved well, and learn how to love, so as to be strong supports in others' lives.

Now cut to Sasha Obama during Inauguration 2009. See Sasha smile, open-mouthed, laughing. We watched as the president-elect walked hand in hand with his younger daughter sweetly bundled in a pink coat and orange scarf wrapped around her neck, contrasting beautifully with her brown skin. The whole world was watching. Half of it seemed to be on the National Mall for the occasion. Up close, there were agents and police with guns, journalists with flashing cameras. But look at her face. Can this girl know fear when she is holding her daddy's hand?

I watched Sasha, and my eyes grew hot and I cried for the umpteenth time that inauguration weekend. I felt as if she were my heart up there on that inaugural stage. If not my heart, the little girl inside me who had never grown up, who, if she could magically manifest herself, could find no better real girl to be.

Sasha and Malia are the nation's favorite "government girls" now. I watch Malia, too, with her camera, snapping her own photographs of the inaugural events, and I see an observant, precocious girl with a keen awareness. Malia is still a child, but she has witnessed the changes in her family, in her life, over the course of the last four years as her father shot to stardom. Watch her, and you know that this brilliant little girl is becoming the kind of teen, much like Chelsea Clinton, who herself arrived in D.C. as a youngster, who is as gracious as she is talented, the kind of daughter that makes the whole world proud.

But it is in such moments that I ache for Malia, and Sasha, too. When will they get to be like their parents, anonymous for years and allowed to make mistakes without the weight of the world's judgment bearing down on them? Fate has handed those sisters this life like a beautiful gift. I am reminded, however, that in German, the word *gift* means "poison." Malia and Sasha live charmed lives, for sure, but every tiara has its burdens, including the public's ideas as to what will be proper behavior for them as they grow older. I remind myself that these girls are just that: young, female human beings. They are not idealized versions of us or our children.

ON CNN AT NIGHT, I hear Paul Begala say that when it comes to choosing candidates, Republicans fall in line but Democrats fall in love. Every time I hear that, my intellect wants to quibble and say, no, that's not true, that any time I supported a candidate there were logical reasons to do so, reasons that had nothing to do with emotion. But my next thought is always, well, isn't the brain good at coming up with rationales for what the heart already wants? To front whatever my subconscious mind thinks will keep me safe?

Paul is right, at least for me: I am a Democrat, and I fall in love with candidates. But it's not puppy love. I want the same qualities in a president that I want in gods and fathers. I fear an Old Testament leader, whose god bleeds wrath and vengeance. Send me a New Testament leader as openhearted as Jesus. I did not grow up with strong, authoritarian parental figures who silenced dissent, where we children had to fall into line, yet I know enough to fear that environment, to fight, so that this land of the free will remain free for people like me and not just the strongmen who have historically hogged the most rights and privileges.

Since 1992, I have looked to Bill Clinton as a father figure, lionizing the man for being as compassionate as he is brilliant. He may have broken promises and vows and overstepped with subordinates, as so many presidents did before him. But this is the power of Bill Clinton: he *still* gives me the feeling that if I were ever in crisis, he would give me, or my cause, a fair hearing. He was not like George W. Bush, who made me feel that he listened only to those in his circle, and that circle was heartbreakingly small.

Then came Barack Obama. Who was this young man who inspired a burgeoning community of supporters to believe that he should be our next president? I had recently begun blogging on the news and opinion Web site The Huffington Post. My very first post on Obama, dated December 1, 2006, was entitled "We Pray for Virgins, Too," a tongue-in-cheek reference to the supposed heavenly reward of suicide bombers. The piece critiqued our hunger for a pretty but blank candidate slate onto which to project all of our hopes and dreams. In those early days, I joined millions of others in our skepticism that the nation was ready for an Obama presidency.

At Christmas 2006, I bought the new hardcover of Barack Obama's *The Audacity of Hope* and borrowed my sister's *Dreams of My Father* and read them back-to-back. Immersing myself in his personal story and the way he discussed difficult issues, including immigration, race, and the economy, I discovered that Obama's way was really the peacemaker's way. Again and again, Obama sat with fighting parties, let everyone speak, and let them know they were heard before bringing the group to thoughtful compromise. He was not "my way or the highway," à la George W. Bush, formerly known as "The Decider." I found myself responding to the way Obama operated.

Hence my problem. As we approached the 2008 primaries, I realized that Barack Obama had won me over. This was no easy state of affairs for a former Bill Clinton staffer, no matter one's experience in the White House. Everything I knew about politics, I knew from the Clintons and their staff. They were my connection to those heady realms of power. But in late 2007, the Democratic Party felt like one big family fight. "You dance with the one who 'brung' you," said one former Hillary Clinton staffer to me as late as October 2008. I countered with "Mrs. Clinton did not 'brung' me, I worked for her husband, and such loyalties don't just automatically transfer—especially when it's for the presidency of the United States." The fact was, in temperament and management style, Bill and Hillary were not mirror images of each other, and the irony was that I supported Obama because, of all the candidates, he was the *most* like Bill Clinton at his best—a leader as compassionate as he was smart, who knew how to synthesize and accept divergent points of view. After eight years of Bush-Cheney, we needed a leader with a supple brilliance who could admit his mistakes. Anyone who smacked of stubbornness made me bristle. Obama's openness felt more revolutionary than his skin color.

At first, my "logical" brain said he couldn't win, that if Americans voted either their hopes or their fears, too many would fear a black president and we couldn't waste this chance to regain power.

But I would also posit that the pendulum had swung so deeply into the dark side with our torture policies, our prisons, and our lack of due process that the exact opposite of the Bush/Cheney administration could look like the Age of Obama. I was torn.

I was not the only one with this problem. In 2008, Rahm Emanuel, now a Chicago congressman, was chairman of the Democratic Caucus, making him the fourth-ranking House Democrat. My former crush had become a powerhouse. Rahm had come from Chicago and was close with Obama and David Axelrod, Obama's top strategist. But the Clintons had definitely "brung" Rahm to the White House dance, and those ties ran deep. For much of the campaign, Rahm officially remained neutral. Then, like so many of us, he declared his allegiance to Obama. So much so that in December 2008, he accepted Obama's offer to be the chief of staff of his incoming administration.

Though I never joined Obama's campaign or the administration, I dreamt about joining. I used to think my government days were behind me. But before us stood an opening into history we could enter with this candidate. Could I join the team as a writer somehow? Or at least do advance work? A few friends sent feelers, but by December 2007, my decision was made: I signed a contract to tell my story with *Government Girl*. I couldn't run off and work sixty- to eighty-hour campaign weeks and finish the book at the same time. But in my blog posts for The Huffington Post, I supported Obama, and repeatedly made his case the best way I knew how. I may not have made a silly YouTube video, but I was an Obama girl through and through.

During the 1992 transition, the Young Turks who had won the presidential campaign—Rahm Emanuel, George Stephanopoulos, and Paul Begala among them—resisted turning to veterans of the Carter administration for guidance. But in the 2008 transition, Clintonites ran much of the show. Former Chief of Staff John Podesta led the transition team. Rahm would be chief of staff.

Mrs. Clinton would become secretary of state. And on the outside, observing and offering guidance for viewers all over the globe, was George at ABC News, still anchoring *This Week with George Stephanopoulos*, the prestigious Sunday-morning public affairs news hour, and reporting on the race during the other ABC news programs. We could watch CNN analyst Paul Begala almost every night, for he was part of the "best political team on television," listening for his pithy takes as he summed up the latest twists and turns of the race. Mark Halperin's The Page was one of the most vital places to turn to for political updates on the Web. Some journalists snarked that for a campaign that promised "change," having so many Clintonites around seemed like "more of the same." But to me, Obama seemed to do what is the mark of ultrasuccessful leaders: he learned not just from his own mistakes but from the mistakes of others, refusing to repeat the Clinton-era blunder of eschewing the White House veterans. Still, I marvel that every power player in *Government Girl* is still in the mix. Some more than ever.

IT'S 2009 and I live in Harlem now. I ran out onto the streets after Obama's victory speech, feeling the kind of sparky community joy I'd known only when hometown sports teams won their championships. As a black woman in America, I had no muscle memory for this kind of political victory. To feel such elation over a presidential win was a new kind of bliss, with cars honking, young people and old people singing and shouting: wow, we were one surprised, happy bunch walking up and down 125th Street. Same for the revelers all across the country who had delivered their candidate to the White House. Together, we had imagined this moment into being. The Obamas carried not just the worries of the world on their shoulders but the weight of our dreams, too.

That's a lot to ask of two human beings.

Sometimes I go to Edmonds' Cafe on Lenox Avenue in Harlem to get macaroni and cheese and green beans to go. Over the counterman's shoulder, taped to the wall, is a color photo of the Obama family. Awe-inspiring, as always. The dream made flesh. But the dream made flat, too. The Obamas made two-dimensional, then reproduced for mass consumption. Step back outside and look up, and you can see the high office building that houses the William J. Clinton Foundation, right there on 55 West 125th Street. I sleep literally two street blocks away from where the president keeps his office—even though this makes no material difference in my life. President Clinton does his work and I do mine, and I've yet to see him, or any staffer I know, on the street below. Yet I find it uncanny that I ended up so near. Walk near Clinton's unmarked office building in 2009, and you will find vendors selling Obama memorabilia. Clinton even hosted Obama in his office—this after the hard-fought 2008 primaries, when in the heat of battle Clinton appeared to compare Obama's candidacy to a "fairy tale" and dismissed the importance of Obama's South Carolina win because the Reverend Jesse Jackson had won there in 1984 and 1988 but had never been within true reach of the party's nomination. I was so angry at President Clinton during the 2008 primaries. I thought he trespassed and spoke words that might have only been political calculations but smacked of meanness. Clinton showed again what we should never forget—that despite his incredible talents, he is only a man. Human. Capable of great desire, including the desire to win at all costs.

Same for the Obamas. They have desires, too. According to the mass-produced calendars, T-shirts, and buttons, he and his wife and children are just flat faces. We love them because they are anything but. If I have one wish for America, it is my hope that when our leaders stumble, as they will, when they hurt others and themselves, which is inevitable, that we will be as compassionate to them as we sense they would be with us if the faults were our own. Our leaders are not gods, and they are not our fathers. But they can be our best hope for peace among nations. They can lead by personal example at home and abroad. I tell these stories of my past experiences with President Clinton, his staff, his agents, and the village of folks who indeed raised me because I think it's important to be honest about human behavior, that it's important to forgive, that it's important not to hold one another to unreasonable standards. That first and foremost we must strive to understand one another. If we do, our hearts can expand enough to accept others for the people they are, not for the flat icons we may wish them to be.

ACKNOWLEDGMENTS

I am grateful to the many mentors, colleagues, family, and friends who helped me write this book.

A heartfelt thank-you to the entire Ecco Press/HarperCollins team. Thank you to Dan Halpern for his support, and to my editor Ginny Smith for her guidance and stewardship. Thank you to copy editor Lynn Anderson for her care with the manuscript, and to Emily Takoudes for her early championing of the project. Thank you to Rachel Bressler, Katharine Baker, Kate Blum, and Beth Silfin for their invaluable assistance. Thank you to my agent, Lisa Bankoff, who shared her wisdom generously and supported me every step of the way.

Alex Steele mentored and befriended me when I was just a stranger in Houston with a fledgling Katrina project. His steady guidance ever since has been an enormous gift to me. Fellow Detroiter Amy Benson welcomed me into her Columbia University workshop and provided close, kind readings of early chapters. Holly Masturzo has been a devoted friend, confidante, editor, and asker of necessary questions. Peter Markus, Hobey Echlin, and Hannah Assadi gave feedback on early chapters, as did Olivia DeCarlo, Joshua Reinhold, Petal Largie, Marguerite Van Cook, and Jacquie Wayans. Thank you to you all.

I am grateful to the following friends and loved ones for helping in ways large and small: Edward Le Melle, Patrice Le Melle, Stephen DiCioccio, Rick Wiener, Trey Ellis, Bill Katovsky, Amed

Khan, Thomas Sayers Ellis, David Wentworth, Brian Alcorn, Megan Moloney, Suzanne de Boisblanc Tyler, Joe Lindley, GDAP, Clem Harris, Max Feldman, Sarah Lewis, Dan Kaufman, Clark Sommers, Stephanie Elizondo Griest, John Mutter, Elyse Jerry Lawson, Shay Youngblood, Larry Blake, Michael Garran, and Glenn Blake. Thank you to those who supported me during and after the time these stories took place: Heather Beckel, George R. Stephanopoulos, Paul Begala, Sidney Blumenthal, Steve Livingston, Gloria Nixon-John, Jeffrey Schox, Jonathan Battye, Justin Leach, Andrew Hutson, AMP, Ralph George, Allen Churchill, Randy Papadopoulos, Stan Buchesky, Bob Myman, Christina Salay, Peggy Podnar Bodin, Susan Armstrong Krawczyk, the Wilson brothers, Stephen Tighe, my InsideOut Literary Arts Project family, including Terry Blackhawk, Kristin Palm, and Robert Fanning, and my Writers in the Schools family, including Robin Reagler, Bao-Long Chu, and Jack McBride. And of course, the entire crew at Empire Café in Houston.

I must especially thank Elizabeth Smith. She allowed me to stay with her in her Mississippi home to write the first chapters of this story. She was ninety years old and a die-hard President Clinton fan with an "I Miss Bill" sticker on the car she still drove. Thank you to her and to her family in Atlanta, including Becky and David Darden, and David's mother, Betty Mac.

Finally, thank you to my family: Carol, Rufino, Elizabeth, and Gladys. Thank you to Kai for so much. You helped me follow my heart.